*Routledge Revivals*

# Transport Network Planning

Originally published in 1979 and with a case-study from Indonesia, this volume examines the question of planning the provision of transport facilities as a special case of the general planning problem. It deals with the modelling (including conceptual short-comings of it), analysis, estimation and control of transport planning and the challenges associated with planning in uncertainty. As well as devoting specific chapters to network planning, the book also provides background material on transport planning, locational theory and economics.

First published in 1979
by Croom Helm Ltd

This edition first published in 2021 by Routledge
2 Park Square, Milton Park, Abingdon, Oxon, OX14 4RN
and by Routledge
605 Third Avenue, New York, NY 10158

*Routledge is an imprint of the Taylor & Francis Group, an informa business*



**Publisher's Note**
The publisher has gone to great lengths to ensure the quality of this reprint but points out that some imperfections in the original copies may be apparent.

**Disclaimer**
The publisher has made every effort to trace copyright holders and welcomes correspondence from those they have been unable to contact.
A Library of Congress record exists at LCCN: 79322642

ISBN 13: 978-1-032-02344-1 (hbk)
ISBN 13: 978-1-003-18299-3 (ebk)

DOI: 10.4324/9781003182993

# Transport Network Planning

Patrick O'Sullivan, Gary D.Holtzclaw and
Gerald Barber

Routledge
Taylor & Francis Group

# TRANSPORT NETWORK PLANNING

PATRICK O'SULLIVAN, GARY D. HOLTZCLAW
AND GERALD BARBER

CROOM HELM LONDON

Croom Helm Ltd, 2-10 St John's Road, London SW11

British Library Cataloguing in Publication Data

O'Sullivan, Patrick
Transport network planning. – (Croom Helm series in geography and environment).
I. Transportation planning
I. Title II. Holtzclaw, Gary D.
III. Barber, Gerald
711'.7 HE193

ISBN 0–85664–742–X

Printed in Great Britain by offset lithography by
Billing & Sons Ltd, Guildford, London and Worcester

# CONTENTS

# ACKNOWLEDGEMENTS

The research underlying this book was supported by a National Science Foundation grant (SOC77-19008). We are especially grateful for the assistance of David Burik in its execution.

# 1 TRANSPORT INVESTMENT AND DEVELOPMENT

## The Geography of Growth

The control or promotion of the expansion of transport facilities is perhaps the most powerful geographically-specific instrument that government can use to guide economic development. That transport is not only important but fundamental in structuring our material existence is suggested by the role which MacKinnon (1978) has attributed to the solution of logistic problems in the evolution of man's adaptability and greater brain capacity. He postulates that it was the need to conserve effort in seeking food which triggered the development of this calculating, pliant creature, man. Similarly the means man has devised to move himself and things from place to place have a profound bearing on the evolution of his institutional and economic arrangements. In examining the history of how man has satisfied his diverse needs and attained his goals through scarce means, it is clear that, whatever the ends, the means are limited by man's political and mechanical capability to manage space. These limits on control over the physical environment have been reduced through time but at an uneven pace. This we can see in the very different levels of material well-being throughout the world. The process of geographical change is visible in the spread of innovations and development. These processes in their turn are accomplished by the permissive provision of means of movement and communication.

Musgrave (1959) has partitioned government's purpose in guiding the economy into three components, a trichotomy of function which may be seen readily in the transport investment process. The elements are:

1. *Resource Allocation Function* The provision of social goods or the process by which the mix of social goods is chosen and by which social and private goods are distributed.
2. *Distribution Function.* The adjustment of the distribution of wealth in conformance with societal notions of fairness and justice.
3. *Stabilisation Function.* The use of budgetary policies to maintain high employment, relative price stability and appropriate rates of economic growth.

DOI: 10.4324/9781003182993-1

The transport sector's share of resources is determined by the necessity of overcoming distance in meeting society's needs. A major determinant of the shape of the economic landscape as it evolves then, is the manner in which distance is conquered. A society at a given state of technology will attempt to develop a transport network to minimise the total disutility of distance. Activities will competitively locate in relation to the network in such a way as to maximise the total utility of the system. Differing needs and abilities to move goods and services will result in arrays of central places and market areas, geographical specialisation in production and the varying scopes of political, cultural and social territories. This differential response to the friction of distance changes over time with technical improvements in transport. Reductions in transport costs enable the drawing of larger markets and resource bases into an integrated economy.

To clarify this contention consider the sale of a single good in an abstract geographical setting. A producer located at 0, sells his product at an ex-works price p, representing the cost of production and normal profit. The demand for the good decreases with price with consumers taking quantity $q_1$ at price p as in Figure 1.1.

Figure 1.1.

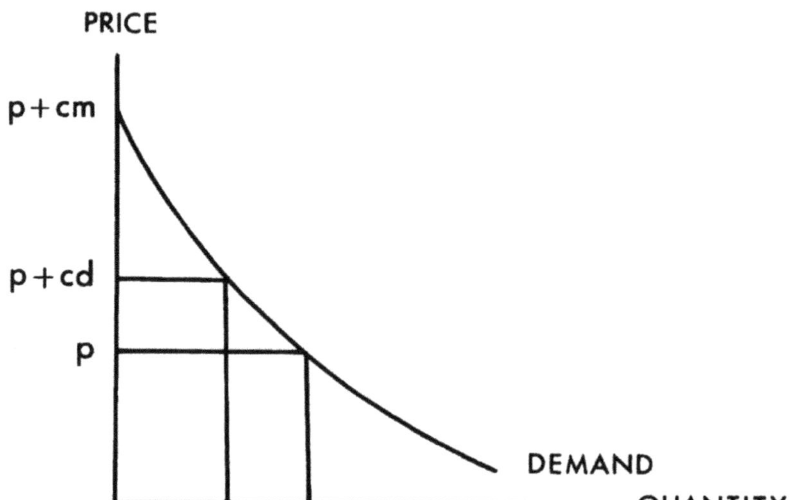

The consumers are located at a uniform density around 0 on an unbounded plain with transport costs increasing equally as a function of distance in all directions. The consumer pays a delivered price of p + cd, where c is the unit cost of transport per mile and d is the distance they are removed from 0. Assuming no scale economies or terminal costs in transport, delivered price will increase linearly with distance from the factory as in Figure 1.2.

Figure 1.2

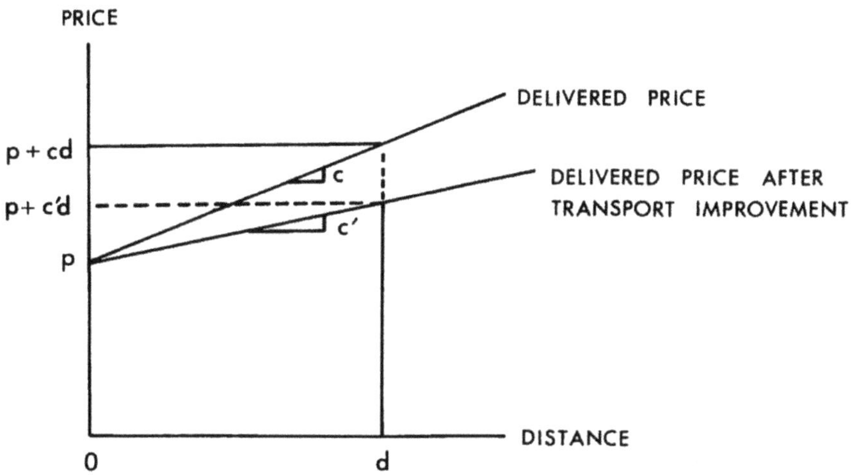


The consumer located at the factory gate will pay p and consume $q_1$, while the consumer at distance d from the factory will pay p + cd and consume $q_2$. The quantity consumed will diminish geographically as delivered price increases with distance from the site of production. In the abstract space we have specified we can describe a geographical demand cone centred on the factory at 0 as in Figure 1.3. The quantity consumed decreases from a maximum at distance zero to zero at the margin of the market m where demand vanishes at prices p + cm on the demand curve of Figure 1.1. If investment in the transport system reduces the unit cost of transport from c to $c'$ and all else remains equal, the delivered price function will rotate down as in Figure 1.2. Delivered prices will be reduced to p + c'd and p + c'm and by c – c' at every location at some remove from the production site. This will induce an increase in demand at every distance from the source and rotate the spatial demand cone upwards and outwards in Figure 1.3. This will extend the margin of the market from m to m' bringing new participants

Figure 1.3

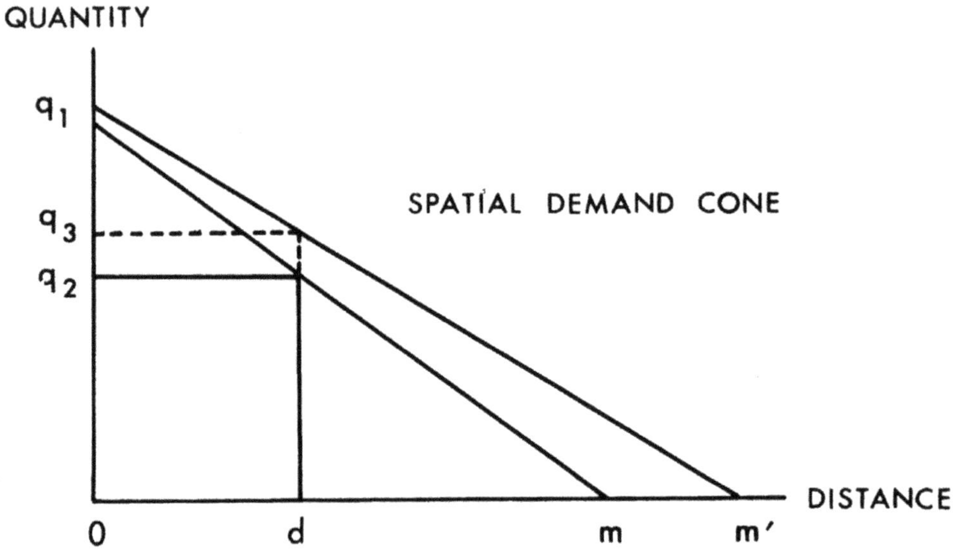


into the economy, as well as expanding the consumption of previous customers, from $q_2$ to $q_3$ at radius d in Figure 1.3, for example.

This analysis can readily be modified to treat the supply of geographically dispersed factors of production, resources or products to a central market or processing point, with similar results in terms of reduced transport costs extending commerce and increasing material well-being. As an illustration we can adapt Walters' (1968) Ellet model from the developing agricultural scene in which he set it to urban industrial circumstances. Consider a city with one employment focus at 0 on a straight coastline, as in Figure 1.4. The circumstances of Chicago on Lake Michigan or Toronto on Lake Ontario are not far removed from this.

The original transport system consists of an infinitely dense square lattice of streets with a uniform cost of travel of c per trip mile. The residents are located in this plan at a uniform density with one worker per residential site. Workers are faced with the choice of subsisting at home at a cost of s per family, just covered by the dole, or travelling to work in 0 at a cost of c per mile for a fixed wage w. The demand for labour at 0 is infinitely elastic at wage w. We can view $w - s + r$ as a location surplus. The margin of the labour shed in these circumstances will be given by:

Figure 1.4

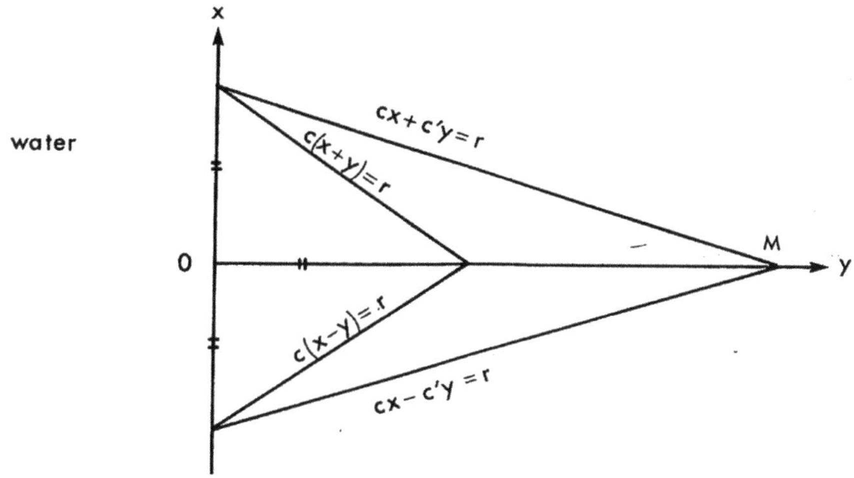


$$c(x + y) \leqslant r, \quad y \geqslant 0, x \geqslant 0$$

and

$$c(x - y) < r, \quad y < 0, x \geqslant 0$$

(From now on we may drop the mirror image expression, remember its existence and operate in the positive quadrant). The rent associated with a location is given by $r - c(x + y)$. This is at a maximum of r at 0, is zero at the margin of the labour shed and totals $r^3/3c^2$. The total number of round trips to the employment site in any day is $(r/c)^2$. It is proposed to build a new motorway straight east from 0 to a distant city. What are the benefits of this improvement? How can they be measured? The investment is beneficial in that it reduces travel times along its alignment reducing the cost per mile from c to $c'$. This makes greater incomes from employment at 0 available to more people and increases the locational surplus of present travellers. The new limit of its labour market will be given by $cx + c'y \leqslant r$. This is more far flung than the original limit. The new labour shed will stretch out furthest along the new road while not extending at all along the coast where no cost reduction for travel is available. At every position away from the coast some advantage can be taken of the new road up to the limit at M. The total number of trips to and from work will increase to $r^2/cc'$ and total rent will increase to $r^3/3cc'$. The total benefits arising from the investment can be measured by the difference between the two rent volumes: $r^3/3cc' - r^3/3rc^2$. The response of travel demand to the reduced cost will be given by:

$$\epsilon = \frac{c}{(r/c)^2} \cdot \frac{r^2/cc' - (r/c)^2}{c' - c}$$

$$= \frac{c^2}{r^2} \cdot \frac{(r^2/c)(c/cc' - c'/cc')}{c' - c}$$

$$= \frac{c[(c - c')/c']}{c' - c}$$

$$= -c/c'$$

and as the difference between $c' - c$ goes to zero so the elasticity of demand for trips with respect to transport costs goes to $-1$. The unitary value for elasticity of this model is obviously an upper limit and relaxation of the various assumptions about the elasticity of demand for labour, the length of the road, the uniformity of transport costs, etc. will squeeze the extent of the margin, reduce the elasticity value and curtail the benefits to be expected.

Devised as a theoretical basis for an operational rule of thumb in the evaluation of agricultural penetration roads in colonial economies, this model can hardly be applied directly to urban problems. The structural device of a single focus of demand to generate neat analytics would obviously be lost in the multiplicity of employment and activity nucleii of a real city. The model does, however, isolate the components of the benefits of transport improvements insightfully, albeit, statically.

With the increase in the potential market for any good, resource or factor resulting from the above process, the possibility exists of an increase in productivity due to the increase of trade, and specialisation of production according to the comparative cost advantages of different areas. This results in the growth of service and manufacturing industries. The expansion of markets may induce the application of improved mechanical methods to production, leading to further increases in productivity and wealth. Increases in the range of markets call for geographical adjustment in the location of production to reap the most from a new set of opportunities. Activities compete for locations according to the advantages they derive from nearness to other activities, markets and resources. Broadly, those enterprises whose viability requires a greater amount of exchange with the rest of the economy will pre-empt more generally accessible locations. Service and manufacturing activities emerge as nodes of localisation at strategic junctures of the network with their geographical frequency governed by their market threshold levels. The concentration of population involved in factor-intensive activites leads to urbanisation of a sector of the population.

The concentration of non-agricultural production to avoid movement costs, generates further growth. Multiplicative effects of development are usually stimulated at existing towns and cities. The economies of agglomeration, which are to a major extent savings on communication and transport, and investment in urban infrastructure induce further growth at urban foci. Traffic is heaviest and increasing most rapidly at these concentrations. Cost savings from investment will be greatest in these sections, encouraging their improvement and further increasing their attractiveness.

Further reductions in the cost of distance with investment and technical advance can again expand the scale of social and economic interaction, extending development into the periphery of the nation. The process of geographical integration involves a gradual penetration of commerce into tradition, with duality reflecting the vestiges of impermeable older ways and inhospitable terrain. The dual face of the landscape negates any simple notion of geographical equilibrium of location patterns. However, the political attachment and intercourse of traditional with commercial regions leads to adjustment through migration, spillover and the continual shrinkage of social and economic distance.

**Transport and Government**

The transport system provides the channels for the trade which sustains growth and an efficient growth path for an economy's geographical structure must be supported by the provision of sufficient transport capacity at efficient prices. The theory of prices and trade between geographically separated markets postulates a set of prices for a good such that inter-area differences do not exceed the marginal cost of transport. This can be demonstrated using Cournot's (1838) two region trade model and the device of back-to-back supply and demand graphs, as modified by Samuelson (1952). Consider trade in a single good which is produced and consumed in two geographically separated markets, regions 1 and 2. We may analyse traffic between them and price adjustments by swinging the quantity axis of region 2 through

180° and placing its supply and demand curves back to back with those for region 1, as in Figure 1.5. Volume of trade translates into the demand for transport so we may project the balance of trade down onto a third graph representing the supply and demand for transport services.

Figure 1.5

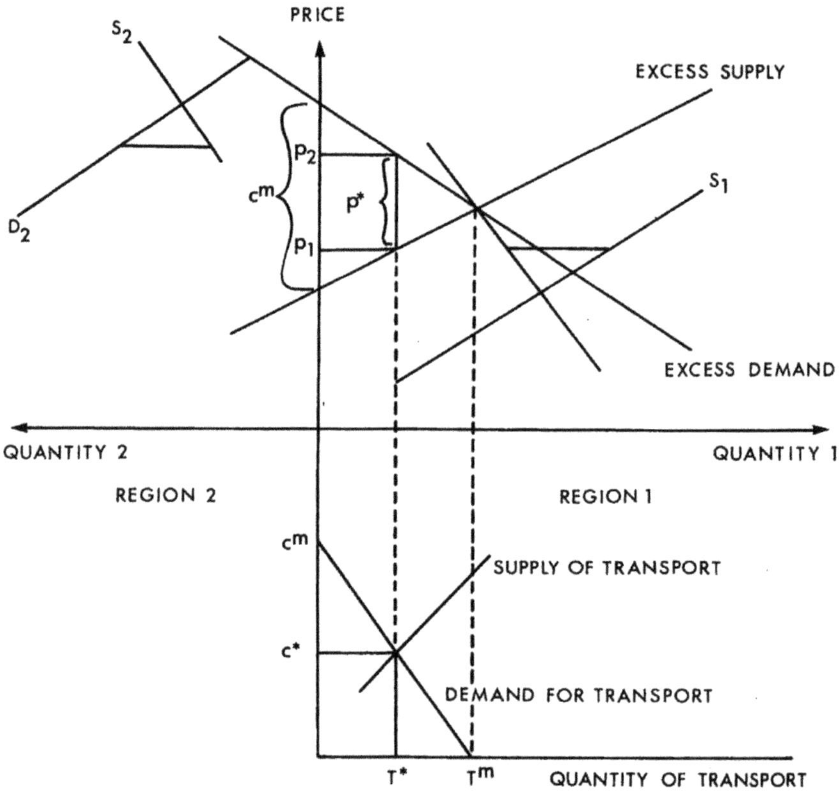


From the supply and demand conditions in the two regions, excess supply and demand curves are constructed by reading off by how much demand exceeds supply or vice versa in each market at every price. The equilibrium level of trade is found where the gap between the excess supply and demand curves just equals the cost of transport. Samuelson showed that this was not only an equilibrium but also an optimum in terms of maximising a measure of social welfare in both markets. The market for transport services is introduced to the picture quite readily. Where transport costs are zero, the amount of trade would be maximal at the intersection of the excess supply and demand curves. This point may be projected down onto the transport quantity/price graph in Figure 1.5 at $T^m$. If transport charges exceeded $c^m$ on the price axis then no trade will occur and the demand for transport will be zero. The demand curve for transport obviously passes through these two points. If we know how the long run marginal cost of transport varies with quantity output we can estimate the supply curve for transport. In this case the equilibrium and optimum for transport $c^*$, along with the quantity of transport demand on the optimal level of interregional trade $T^*$ and the equilibrium prices for the good in both regions $p_1$ and $p_2$ can be determined simultaneously. It is clear from this construction that if freight rates charged exceed marginal transport cost, socially advantageous trade will be suppressed. Rates below marginal cost will engender trade in excess of efficient levels. Both eventualities will deter growth.

The nub of transport policy is seeking to ensure the maintenance and development of adequate means of carrying social intercourse and commerce within and without the nation, counting the cost of doing so in terms of political discontent, possible damage to the well-being of other than the immediate users, as well as of the resources of land, labour, machines, structures and materials used. The nation's cohesiveness and health depend on the density and continuity of social interaction and trade within its bounds, on ease of access to markets, and on the protection of its interests beyond its shores. If we accept this proposition as axiomatic, then it follows that, in order to sustain and deepen national cohesion, it is necessary to provide and keep open channels for contact and exchange between the nation's constituents and the rest of the world. Thus, transport facilities and services have seldom been provided without the involvement of government and every transport investment and pricing decision becomes a potential bone of political contention. The policy maker has to decide whether resources should be allocated to particular forms and means of transport on the basis of the market or some administrative pseudo-market process, making choices on the basis of a predetermined rubric. When the intangible and political considerations involved transcend the ability of the invisible hand or cold figuring to comprehend, it may require the proximate calculus and will of statesmanship to resolve the choice and the majesty of the law to enforce it once made. This political resolution can be a one-off investment decision, the designation of a standard, a regulation or a prohibition. Recognising that it is oversimple, never the less, it may aid our understanding if we examine issues in terms of whether the market, administration or politics is the predominant process involved in their resolution.

From an economic viewpoint many transport services display characteristics which deny the efficacy of a market solution in determining the level of their supply. The long gestation and sheer magnitude of investment in the way and terminals of transport systems, have often put them beyond the scope of unsupported commercial interest. There is an ingrained popular view of roads as a 'public good' which should be provided and paid for collectively, and this attitude is not without basis. The cost of operating a place- and time-specific price mechanism and excluding non-payers throughout the road network would almost certainly exceed the additional well-being resulting from a closer equation of individual use of roads and the costs arising solely from their use.

It is asserted that some forms of transport, because of their enormous initial capital costs and relatively small unit costs of running, enjoy a continually decreasing unit cost as their output expands, and thus, constitute natural monopolies. These require strict government control of prices or output to ensure that this power is not abused at the expense of national efficiency. This was the basis for the regulation of railways which inhibited their flexibility and enterprise well into an era in which they faced stringent competition from lorries and cars. A peculiar convolution of the monopoly control argument, which is powerfully wielded in the UK, holds that competition can be wasteful inasmuch as it inhibits the achievement of inherent monopoly advantages, variously identified as 'coordination' and 'integration'. Additionally, competition is held to be an unstable, and therefore, undesirable state of affairs. The prescription arising from these points is that public monopolies be created by nationalisation, or local authority ownership, for the controlled enjoyment of stability and increasing returns to scale. This remedy was applied to public transport in London, and now elsewhere, with little evidence of success to date and little hope for the future. Some socialists would extend the remedy to the whole transport sector, given the strategic significance of the transport matrix in lacing the economy together, which demands public control.

A further reason for dissatisfaction with the ability of an unbridled market solution to determine the right quantity of transport is the widespread external effects which transport generates – penetrating the dwelling and working space of people not travelling with noise and nuisance which the users are not obliged to compensate for in the price they pay.

Natural monopolies subject to collective control are characteristically beset by difficulties arising from underinvestment and inflexibility, as a bureaucratic rubric replaces market response as the key to decisions. In the transport sector many of these problems can be bypassed if the government provides the lumpy, fixed network facilities and the private sector provides the vehicles and services on a competitive basis. This is the usual arrangement with road and water transport. It has not been found feasible with railways which seem to suffer from innate inflexibility and lack of response to changing demands.

Government control of transport may, however, lead to geographical distortions caused by excessive investment in facilities. Freight rates may not reflect long-run marginal costs. This is the case with barge traffic in the USA, where the Corps of Engineers provide costly dams, locks and channels virtually without charge to the user. In such instances the choice of manufacturing locations and agricultural products will not reflect the true cost of resources employed in transport. In the case of manufactures, such distortions are less easy to correct than most. Buildings and machinery are not readily moved and geographical momentum can soon pick up. Badly sited factories are in place for their lifetime and may attract further ill-advised investment in infrastructure and linked activities to their vicinity. The correction of such geographical misdirection of development when detected, can be achieved by instituting proper transport pricing. It should be undertaken to prevent the compounding of geographical inefficiency generating a wasteful burden of movement and trade. Steps such as the removal of cross subsidised branch railway services may cause considerable disruption in the short run. Such pains of transition can, however, be mitigated by temporary subsidies, to replacement road hauliers for example.

## Transport Networks and Growth

Transport networks have in the past been thrown out in calculated, strategic fashion with the objective of controlling territory by the expeditious movement of troops, administrators and information. The state's integrity is ensured by control which is exerted over routes leading to every corner of the realm and providing for the rapid connection of all quarters to the focus of authority. The Roman mesh served such a purpose, as did the railways of the Raj. The fragmentation of power which followed the Roman hegemony in Europe was accompanied by a fracturing of the transport system. The medieval world consisted of local foci for clustered routes serving local needs, sparsely connected with the rest of the world. A slender thread of routes carrying religious administrators and merchants stitched Christendom together. As nation states emerged, so national networks with strategic and administrative objectives were built. In Britain this was a somewhat haphazard endeavour. The more conscious process of centralisation in France created a bureaucracy to plan and administer the routes which focused the state on Paris.

As economies commercialised, the growing transport needs of expanding trade were met by navigational improvements and turnpike enterprises, closely tailored to locally determined requirements. The canal network evolved by similar private corporate action, enhancing the natural waterways, providing the channels for trade to support commercial agriculture and industrialisation. The government's role was essentially one of granting permission.

**British Industrialisation**

British road, river and canal improvements of the sixteenth and seventeenth century laid the groundwork for economic expansion and the achievement of a sustained increase in productivity. This enabled production to outstrip the growth of population, overwhelming the Malthusian spectre, and achieve a real increase in well-being in the 1780s and 1790s. All the more important improvements in transport infrastructure took place before the concentration of production in factories became general. Roads and canals were in place to permit growth. By 1775 the main road network was laid out. Between 1775 and 1800 the arteries of the waterway system were established. The overland transport of exports of cotton, wool, copper, brass and iron to the ports of London, Liverpool, Glasgow and Hull which surged in the 1750s, placed a stress on the rivers and roads of the country. This encouraged the building of turnpikes and canals. Once in place, these facilitated economic advance through industrialisation. They were prerequisites rather than by-products of the process. The canals were the vital element. With a 12:1 advantage in efficiency over wheeled vehicles, barges on canals enormously increased transport productivity. This allowed tremendous increases in efficiency in industries which employed them for carriage. The crucial contribution of canals, however, lay in overcoming the fuel bottleneck of the time. They made cheap and abundant supplies of coal for direct heating and steam power available. In addition to multiplying employment and income, the canals tied together and enhanced the value of the river navigations and complemented and protected the roads by drawing off heavy, destructive traffic. Canal building provided some valuable early lessons in enterprise, engineering, logistics and technology for major works of the future. Britain's leadership in terms of political integration into a nation state under the Tudors and in the emergence of science, commerce, farming and manufacturing undoubtedly owes much to the ease of movement geographically around an entity of manageable proportions for the forerunners of railway technology.

When the rail web was spun in the 1830s and 1840s, the process of industrialisation was well under way. In the British Isles railways were primarily built to serve existing traffic after a careful assessment of the existing road and water traffic. Railways did little in the way of geographical pioneering, largely serving the existing settlement pattern. The railways' coal requirements for the manufacture of equipment and rails and for raising steam were of marginal significance to the country's energy economy. Railway needs did create a new branch of mechanical engineering but much of its output was for export markets. The call for iron for rails amounted to nearly a third of the total demand in the period 1844 to 1851, accounting for most of the increase in its production during that time. By 1851, most of the network was laid down and the demand for pig iron for rails was of little overall importance, especially since recycling of scrapped rails began to have its effect on material supply. The construction of railways was, however, of strategic importance in determining the magnitude and tempo of economic activity in the 1840s. In 1847 the wage bill of the quarter of a million men employed in cutting, filling, tunnelling, bridging and building stations, yards and sheds accounted for 3 per cent of the national income, the total expenditure on railway formation amounting to 6.7 per cent of national income. This generated a pronounced surge of spending which multiplied itself in rounds of income and spending increases through the economy. Railway building combined with an upswing in house building, probably imparted one of the underlying long term fluctuations characteristic of the fortunes of the economy since that time.

**The US Railway Experience**

Fogel (1964) has doubted that railway construction had a profound effect in the USA on the total level of activity and growth achieved, even though it clearly shaped much of the economy's geography. David (1969) countered this with calculations suggesting that if the railroads were treated as one big investment project, they returned 15 per cent to 20 per cent in the way of social benefits in both the 1860s and

1890s. A more relevant measure of returns to marginal additions to the system seems to have fallen to the order of 6 per cent to 17 per cent by 1890. This was roughly in accord with the marginal rate of return on the aggregate US capital stock of about 11 per cent. So by 1890, railroad investment was not any more attractive than the run of opportunities. However, the returns achieved prior to this while the network was building up must have been of great social worth.

In geographical terms, although they played a role in determining the outcome of the Civil War, railheads did not outstrip the settlement frontier till the 1870s. The inhospitality of the Great Plains held the line along the Missouri from the 1840s to the 1870s. The Plains remained an area of passage and pasture until the tracks of the Union Pacific, Northern Pacific, Southern Pacific and Santa Fe spanned their expanse, creating a grainland economy and human landscape. The heavy rainfall of 1875 to 1880 contributed to the success of wheat farming and rail-led settlement, pushing out open range graziers. A drought in 1886 was a reminder of the semi-arid nature of the area and caused a contraction of settlement. In this 'land of risk' the frontier of cropping ebbs and flows with the price of grain and the amount of rainfall, with the cattle industry taking up the slack. The tidemarks of arable cultivation and settlement had, however, extended far enough so that in 1890, the Bureau of the Census declared the frontier closed.

In this process, the political objective of manifesting the nation's destiny interposed itself with commercial concern. Land was given to railroads for right of way and to generate finance for construction, to promote the spanning of the Plains and Cordillera. The considerations involved were those of consolidating a nation as much as those of profitability. The sequence of construction reflected the dominance and victory of northern industrial interests over southern agrarian ones. In the history of railway building, analysis and accounting came far behind laying track and it is no wonder that much of that track ended up as twin streaks of rust across the prairie.

O'Brien (1977) has summarised and evaluated the econometric contributions of Fishlow (1965) Fogel and David to the historical debate over the significance of the railroads in the growth of the US economy, along with those of Hawke (1970) on British, Metzer (1973) on Russian and Fenoaltea (1972) on Italian experience. From this it is clear that the magnitude of railway impacts was insufficient to sustain the 'leading sector' role which Rostow (1960) accorded the innovation. In reducing transport costs, extending markets and inducing geographical specialisation, steel rails and steam locomotion merely extended the process which the preceding 'adventure in hydraulics' had triggered. The new quantitative economic history throws up the conjecture that for Western Europe, Russia and the USA, river navigations and canals made a greater contribution to long-term economic progress and that the greatest contribution of railways was in providing a substitute for costly road haulage.

**Soviet Transport Policy**

Based on these historical experiences, the Western tradition put low-cost transport in the vanguard of the growth process. The rise of Britain, the expansion of Western Europe and the growth of the USA can all be ascribed in large degree to innovations in transport – both maritime and overland. A myth that arose in response to such observations was that the provision of low-cost transport capacity was both a necessary and sufficient condition for economic development. This view held sway in Tsarist Russia and the lavish laying of rails was viewed as the mainstem of the drive towards industrialisation.

In 1917 there was a change of management and of conception of the process. From the mid 1920s, the Soviet government relegated transport to secondary rank in the economy. Its source of investment funds was cut drastically in favour of heavy industry. Transport networks and rolling stock have only been expanded in response to actual or impending traffic bottlenecks. Few projects have been built ahead of demand to stimulate development. Resources were diverted into transport capacity only when this was clearly necessary for expanded industrial production.

Soviet performance has shown that such frugality in transport and rapid industrialisation can be compatible, at a cost. There is obvious evidence of an overbalancing of cost reductions in the transport by time and disruption costs incident on shippers and receivers, especially during the turmoil of 1930 to 1934. More transport capacity would certainly have benefited industrial growth then, and greater generosity in this sector would probably have more than paid for itself since in avoiding waste and inefficiency.

In 1974 the announcement of the start of construction of the Baykal-Amur Mainline, however, evidenced a return to the older model and nineteenth-century capitalist plans for the exploitation of the Soviet Far East by means of a 2,000 mile line parallel to the Trans Siberian, skirting north of Lake Baykal.

**Highway Networks**

The preoccupation with railway building of the nineteenth century left most nations with pitiful road systems. The building of high quality road networks has been a public response to political pressure from a growing number of road users and from the military and construction industry. In the USA there were 1,000 trucks in 1900 and 4,000 cars. By 1920 there were over 1 million trucks and 2 million cars. There is an obvious political incentive to cater to the needs of their owners. The geographical process which accompanied this growing pressure for road space was the explosion in size of major cities drawing in the surplus labour of rapidly mechanising agriculture. By 1920 there were more people in towns than in the country and the major agglomerations continued to grow apace, spreading suburbs served by rail lines and the Model T.

The feeding, clothing, heating and housing of this burgeoning urban population, the exchange of their specialised products and their own travel, generated a massive growth of trade and traffic. To meet these needs there was an enormous building effort in the two decades from 1920 to 1940, financed in part by the imposition of user changes. The road programme was a response to political pressure and, given the over-representation of rural interests in the US political system, it is not surprising that there was a serious imbalance in favour of inter-city and rural roads as opposed to urban transport. Prestige projects appear more magnificent per dollar in a rural setting and are easier to design and execute. Unnecessary roads were built and more were built to excessively high standards for the use to which they were put. In 1950, 148,000 miles of road in the USA were providing no essential traffic function.

The pressure for road space reasserted itself after World War II as car ownership diffused through the society, approaching the saturation level in the late 1950s. The taste for low density suburbs, encouraged by the tax structure and Veterans Administration and Federal Housing Administration loans, engendered a tremendous growth of car use. From 1947 to 1957 vehicle registrations doubled from 34 million to 68 million and miles of travel rose from 340 billion to 650 billion vehicle miles. By the mid 1950s the existing roads were unable to handle the increased numbers of motor vehicles. Bad surfaces, increasing congestion and climbing accident rates stoked the discontent which the road lobby focused on the Congress. The Federal Highway Act of 1956 designated a 'national System of Interstate Highways . . . located as to connect by routes as direct and practicable, the principle metropolitan areas, cities and industrial centers'. It was projected that the 35,000 miles of limited access, intercity routes would improve speeds and safety and that the 6,000 miles of urban routes would reduce downtown congestion, reducing commuter stress and strain. The system was initiated with extravagantly optimistic predictions of benefits, under pressure from an enormous surge of demand, almost totally immune from close scrutiny, never mind detailed appraisal.

Friedlaender's (1965) *ex post* evaluation of the Interstate System suggests that, treated *in toto*, the network produces a positive net present value; however, it is far from being the best possible investment of funds in transport. In many rural areas, the likely level of traffic is too small to support interstate standard roads. Because no minimum standards were imposed on the Interstate System and because the states bear only 10 per cent of the total cost, state highway departments had little incentive to ensure economy and efficiency. Highways were laid down far in excess of any conceivable need. Two lane roads would have achieved essentially the same operating costs as four lanes in many cases. The cost of the whole system could have been considerably lowered without any appreciable reduction in benefits. This does not hold for the urban system, which does seem to have met a strong pent-up demand. It is possible to argue, however, that the marginal cost pricing of urban road space might well reduce congestion and increase benefits.

Promboin (1971) evaluated the geographical impact of the system. He could not reject the possibility that the stock of interstate highways has had no effect on local economies on the basis of state level statistics. Much of the country, including the midwest heartland, appeared to be totally unaffected by the construction of the system or the additional capacity it provided. In the mountain states, with sparse populations, where enormous lengths of road were laid down to span its extent, the system did have a noticeable effect on employment. It was only in the southeast that the capacity of the system seems to have had a positive effect, overcoming a bottleneck due to limited capacity.

For a geographically detailed investigation of the developmental effects of highway investment we turn to the UK motorway experience. It was government policy 'to ensure that . . . the road programme should reflect regional development needs; and careful account was taken of the benefits which new or improved roads could bring to less prosperous and more remote regions' (Department of the Environment 1972a). These benefits are observable either as savings to increased traffic or in the increased activity which generates it. The difficult task is isolating the increase solely attributable to road building. In a study of the effects of the Trans-Pennine Motorway (M62), Dodgson (1974) could detect only a very small effect. The impact of this road would only reduce total costs of manufacturing by a maximum of 0.3 per cent in its close proximity and less elsewhere. Some slight relation between employment growth and transport cost reductions was discernible but this generated very small numbers.

It does seem that the effectiveness of highway building in stimulating regional growth is limited and uncertain in developed economies. It is quite possible that reducing transport costs to the undeveloped, peripheral areas of an economy may have the reverse of the desired effect. For a product which enjoys decreasing costs and which is presently produced both in the centre of the economy and on the fringe, protected there from competition by high transport cost, a lowering of transport cost may result in greater concentration at the centre and decline in the fringe as the peripheral market is penetrated by central producers. In addition, much of the provision of interstate and motorway capacity is in excess of requirements, especially if speeds are to be limited to 55 mph as in the USA. Much of the interstates' vaunted benefit stemmed from their design speeds of 70 mph.

These reservations should not mask the fact that in many countries, where transport facilities are inadequate, lack of roads may present a serious deterrent to regional and national growth. Indeed, transport investment could be considered a necessary cause in economic development since the potential for exchange can only be realised when the channels for its movement exist. Clark and Haswell (1964) put it aptly:

> It cannot be emphasized too strongly that the first requisite for the improvement of production of a subsistence economy is the provision of transport . . . If there is to be any improvement in an agriculture which at present only provides for subsistence, crops must be grown which can be sold. Selling implies transport to a market. The methods of transportation available to a subsistence agricultural population . . . are so costly that produce only has to be carried a limited distance before most of its value from the point of view of net returns from the sale, has gone.

What goes for crops, goes for manufactures and services as well.

## Conclusion

We have seen that the provision of transport facilities, particularly the road network, is bound to involve government and collective choice. Although the impacts of latter day additions to a massive existing structure of road capacity in developed lands may not be enormous and crucial to national well-being, the building of the right components at the right time can have a major effect on the trajectory of national productivity. This is especially true in the early stages of the emergence of a commercial economy, or of the growth of a city, at the more local scale. The wrong network structure for the needs of a particular stage of development can badly distort the geographical evolution of a nation or a city's spatial economy. Once set in motion, an inefficient locational structure may perpetuate and amplify the waste due to the operation of geographical inertia.

To plan well and provide the right network capacity at the right time is important to national welfare.

# 2 THE PRACTICE OF TRANSPORT PLANNING

## The Objective

In his essay *The Road*, Hillaire Belloc (1923) gave as an objective for transport planning:

> To find a formula of minimum expense in energy for communication between . . . given geographical points under given conditions of travel and carriage.

When we look about us, we see a landscape littered with the detritus of the failure to forsee adequately the problems of achieving this objective. Whether the burden of guilt lies in a misformulation of properly foreseen givens or lies in misforeseen givens that were properly formulated is not always easily ascertainable amid the conflicting *a posteriori* evaluations of the investment decision. Irrespective of the source of failure, these artefacts of well-intentioned planning do register the real cost of ignoring or misjudging the uncertainty which pervades our knowledge of the world and its future. Mismatches of needs and capacity are the wages of inflexibility. The geographical imbalances of demand and supply, visible in congestion and in surplus and abandoned capacity, arise from improperly considered costs and technical changes, changes in the geography and magnitude of demand and changes in the tastes and preferences of consumers and the body politic.

The realities of *a priori* uncertainty have shown themselves as quite sudden truncations of the continuity of expectations. The history of transport is replete with radical switches between means of movement, as technical advance provided a threshold of economic advantage to a new enterprise. Railways superseded canals and in turn succumbed to the competition of trucks, cars, and aircraft; silted channels and overgrown tracks are the marks on the land of these discontinuities. More recently we have the spectacle of unused stubs of expressway in Milwaukee and Boston, for example. More dramatically, yet less perceptibly, the massive mileage of lightly trafficked interstates span vast open space, bears witness to an imbalance between provision and need. Slashes of empty divided highway connecting Port-au-Prince and Accra to tropical nowheres show that such waste is not a luxury of developed lands alone. The twice daily metropolitan queues for road space and


DOI: 10.4324/9781003182993-2

train seats evidence shortfalls of capacity. The effects of transport sector misjudgements are not confined to appearances of mouldering or straining routeways and terminals. The effects can be more widespread sectorally and geographically: the limp order books of Boeing and Lockheed and the disruptive deceleration of Californian growth in the early 1970s owe much to mistaken foresight in airline fleet planning. In memory of this, the new surge of demand of the late 1970s is being met with greater caution and a geographically more widespread production stance.

There is clearly a need for network development decision methods which reach for desired goals in such a fashion as to reduce the potential burdens of uncertainty in the specification of future givens and the concomitant formulation of an objective. The aim of this volume is to devise and analyse an approach to solving this need. We concentrate on choices of route additions, since these are locationally specific and require geographical insight for their success. In particular, we address the institutionally collective provision of road capacity, where a broad definition of public good must be employed and a correspondingly greater burden of doubt must be borne.

The doubts which shadow a decision firstly concern our understanding of the workings of the world and, thus, how it will evolve in space and time. The second source of unease concerns the stability of the preferences between various goods which the affected population do and will have. In planning practice, these elements translate into uncertainties about costs, demand and utility. Some examples of these uncertainties may clarify the concern in the transport context.

### Cost Instability

Changes in transport operating and track costs, the balance between them and the advantage of different means of transport for different tasks may result from changes in the relative prices of inputs. Of specific interest to transport planning, is the fact that, with depleting reserves, prices of oil-derivative fuels may well be expected to increase in relation to other inputs of production and consumption activities. Of general interest to planning, cost changes may accompany the realisation of the falsity of past experience.

A striking example of this re-evaluation is to be seen in the specification of road capacities in the UK. In a paper addressing the issue of uncertainty in road investment appraisal, Quarmby (1967a), then a Ministry of Transport civil servant, suggested that: 'There is not likely to be a radical change in the vehicle design or track requirements within fourteen years from now sufficiently large to cause design standards for roads to be changed . . . thus . . . there is likely to be much greater uncertainty in benefit estimation than in cost estimation.' Seven years later the agency issued Technical Memorandum H/6/74 (Department of the Environment, 1974) with new design standards for inter-urban roads and high quality urban roads, because experience had shown that greater volumes could be handled than allowed for in the previous standards:

> One effect of the new standards is to increase the average daily flow by up to 50% for dual carriageway roads and 100% for single carriageway roads where grade separation is used.

This clearly reduces the cost of providing roads for a given demand considerably. *Sic transit lex scripta.*

A decline in the cost of sea transport relative to land carriage with container, roll-on-roll-off and barge-shedding ships has made much port capacity redundant, decentralising the points of land–sea egress and overland flows of traffic. These developments could well smother the prospective advantages of land transport connected schemes, such as the Anglo-French Channel Tunnel. Though shelved for political reasons, the tunnel's potential economic value is evaporating rapidly with the technological improvement of competitive means of transport.

**Misapprehensions of Demand**

The demand for transport by a particular means or in a particular place may surge or plummet quite at odds with naive extrapolations of current trends. The surge of demand for coal movement in Britain's industrial revolution swamped the capacity of the natural waterways that served as her veins up to the early eighteenth century. Expectancy of exponential growth in air travel proved mistaken when it turned out to follow a logistic path, leaving US airlines with an excess of capacity in the mid 1970s. The same trend in international travel was foreseen early enough to be brought to bear in suspending commitment to a third airport for London. Now in 1978, the air travel industry faces an unexpected increase of demand due to price reductions inspired by Freddie Laker's enterprise and Alfred Kahn's chairmanship of the Civil Aeronautics Board.

Revealing examples of patently false hopes can be seen where decisions are decentralised to local governments and needed synergistic benefits fall prey to political competition. For example, in the UK, the

the sum of local authorities' expectations as to central area office employment and, thus, commuter volumes, far exceeds any reasonable estimate of total national employment in this category. These excessive prospects then fuel local bids for central government funds to subsidise rail services or increase road capacity.

**Evolving Tastes, Fashion and Welfare**

Perhaps the most perplexing component of insecurity in making choices now for future satisfaction lies in the worth potential consumers attach to various goods and bads. The calculation of preferences eventually comes down to the specification of alternative decisions and their outcomes and the construction of an objective criterion with which to weigh the balance of costs and needs implied by various courses of action. The problem involves the construction of the social welfare function. More specifically it involves the determination of the present preferences among different things, including present things and future things, and the presently forecast future preferences among equivalent things. The complication involves not only the specification of time-preferences of the society that commits to the investment but also the values and norms of the future societies that will be required to use, maintain, and expand upon the initial investments. Since each of these activities are to some extent constrained by present commitments, viable future value systems are constrained accordingly.

In road planning, a switch in the 1960s from engineering concepts of efficiency to the economist's maximisation of consumer surplus did much to change the accent of solutions. The achievement of performance standards was replaced by the balancing of a discounted stream of generalised cost of travel savings against the present value of construction and maintenance costs. As the economising of expenditures and time gave way in weight to collective environmental desiderata, so the objective of the planning exercise and, thus, good design further changed.

With a deeper appreciation of the interwoven relations between land use and transport, there was acceptance of a more all-embracing appraisal of potential lost opportunities and the soundness of more simply motivated choices became questionable. These matters of selection criteria are subject to the quirks of fashion, especially as people respond to political dramatising, hence there is an uncomfortable instability with respect to the necessarily lengthy planning horizons of transport systems. For example, the 'environmental' goods, whose value rose throughout the 1960s, in comparison with that of the ease-of-travel objective of the 1950s, suffered an erosion of worth in favour of energy efficiency after 1973. These changes in objectives were major and all within the planning horizons of the grandiose transportation studies of the late 1950s and early 1960s.

These historical instances of evolving consciousness and changing sensibilities should serve to provoke care in the specification of the goal of transport investment.

**Transport Planning in Practice**

As a preamble to proposing ways of recognising and accommodating uncertainty in transport network investment and design decisions, it is instructive to examine the planning process by considering how decisions have been made and how the future was anticipated and characterised.

The track which cars and trucks began to fill in the early twentieth century was traditionally a collective good. The brief effort at marketing road capacity in the turnpike era was not commercially successful. Thus pressure of demand was met with a slower and governmentally controlled expansion of roads for motor vehicles. The concentration of power in nations combined Keynesian expenditure with military motives to engineer autostrada, autobahnen and interstates. These were designed to link the centres of population concentration at the highest engineering performance standards, almost regardless of geographical demand variations and any relation between cost and traditional notions of social utility. The explicitly military justification offered for such programmes silenced critical appraisal in times of intense chauvinism.

As ardour for road constructions cooled, more careful analysis of the worth of increments to the transport system was employed, inspired by the experience of cost-benefit appraisal of water resource projects. Such analysis did not come to bear on the marginally valuable elements of the system, which had been built first because they were rural, easier to build, and more visible per unit cost. The elements of the system subjected to close scrutiny were the difficult urban legs, which Friedlaender (1965) has shown were the most beneficial in travel cost terms.

The theoretical framework for such appraisal of transport projects had been laid down by Dupuit (1844) and Ellet (1839) in the context of bridges, roads and canals. Drawing on his personal experience in the construction of Mexican railroads, Wellington (1887) contributed to the engineering and geographical specifications of the problem. There is a largely ignored example of public, economic evaluation of an entire network dating from the 1830s. The Drummond Commission's (UK 1838) proposals for an Irish rail network were based on estimates of demand taken from population and current traffic surveys, construction costs based on a national geological survey and operating costs from the Liverpool–Manchester line. This model of analysis awaited re-invention and elaboration over a century later under pressure of demand for urban road space. In the 1950s, a mounting awareness of the interrelatedness of land use and transport found expression in the work of Mitchell and Rapkin (1954) and the economics and operations of transport systems were given clear formal statement by Beckmann, McGuire and Winsten (1956). From this there emerged a generally accepted corpus of methods and models used in urban land use and transport planning. The purpose was to ascertain current movement patterns and their relationships to the geographical distribution of activities and the media of movement in an attempt to predict future requirements for movement, which could then be used to design an economically and socially acceptable circulation system. The best way to explain the concepts and techniques involved is to trace the evolution of the creation.

### The Urban Land-use/Transportation Study Models

The counting of traffic volumes on various sections of a network as an aid to design has a long history and counting techniques are still of utility for more local design problems, for inter-urban and rural road planning and for keeping a running check on trends in traffic. Flow patterns, however, tell little about specific requirements for movement except how traffic uses the existing network. In the USA in the 1930s and 40s 'origin/destination' survey techniques were developed to detect more precisely what desires for movement were, to build up 'desire-line' patterns linking places between which people desired to travel directly. Initially the surveying was done at road blocks and all vehicle users were interviewed. This was expensive and led to congestion which was partially alleviated by taking a sample of vehicles. The improvement in questionnaire design and sampling techniques enabled the introduction of household, home interview, sample surveys of travel requirements. Desire line maps produced from these surveys were then used by engineers to 'eyeball' reasonable network developments.

To augment the origin/destination home interview data, roadside counts were taken at an 'external cordon line' thrown round the area under study to catch traffic entering and leaving it. Questionnaire surveys of firms provided information on commercial vehicles, bus and taxi movements. As a check on the sample expansion factor used in expanding household data from the sample rate to the universe of households, a 'screen line count' was instituted. This provided a check on possible under-reporting of trips. At a line thrown across the area, preferably where there were few cross routes, e.g. at a river or railway line, a traffic count was made. If the counted traffic exceeded the amount calculated from the sample survey, the interview sample was expanded by a proportionally greater ratio. There is usually little under-reporting of journeys to work but social and recreational trips are usually understated in household surveys, so trip purposes were established for the screen line count and differential expansion factors can be used for different trip purposes in an attempt to equate trip estimates from the household survey crossing the screen line with actual trips crossing.

A major turning point in the development of transport studies occurred in 1954 with the publication of Mitchell and Rapkin's *Urban Traffic: A Function of Land Use*. The obvious relationship between traffic flows and land uses had often been remarked before, but Mitchell and Rapkin, in stating the idea that the physical interchange of people and goods was *measurably* related to activities on the land, isolated the elements of a model of urban behaviour. However, they did not suggest how the model could be made operational to help in the urban planning process. Also in 1954, Voorhees stated the possibility of using the 'gravity model' hypothesis about human spatial interaction for the analysis of urban traffic. It was also in 1954 that Douglas Carroll Jr completed his study of Grand Rapids, which exhibited his expertise in questionnaire design, data handling, and the supervision of extensive urban surveys. On the basis of this, Carroll was invited to take charge of the Detroit Area Transportation Study. This provided the first opportunity for the application of Mitchell and Rapkin's and Voorhees' concepts and resulted in the first large-scale operation of a 'traffic model'. Using origin/destination data from a 3 per cent household sample and a gravity model for distributing trips between zones, 1970 traffic flow patterns were simulated from 1970 land-use projections.

The Detroit study used the regional and local land-use plans for 1970, which had been arrived at by the 'drawing board' allocation of activities and population to land in the area on a vague circulation system skeleton. Up till the 1960s the method of producing a land-use plan remained an outgrowth of physical designers' techniques as elaborated by town planners since the 1900s. Roughly, the method consists of estimating population for a future date, usually a ceiling population for twenty years ahead. The requirement then, is to provide a final urban form for this 'ultimate' development. The population projection is translated into a number of households which are allocated at some design standard average densities. Acreage for residential and service requirements at the ultimate date is set aside. The existing amount of development is subtracted to give an increment to be distributed geographically. The number of jobs necessary to employ the population and the increment over current employment are found. Land for commerical and industrial uses necessary to employ the increment is allocated at some given design standard densities. The current circulation system is mapped and extended to meet the requirements of some board, generalised concept of 'desirable urban form' (seen from a bird's eye view and with the architectual aesthete's vision). The incremental land uses are allocated around this framework and so by intuition and eyeball techniques a plan is arrived at which haphazardly compromises the subjectively valued, most desirable attributes of a city from the planner's viewpoint. This was the method used in Detroit, and it unfortunately still holds sway in many places.

Carroll moved on to the Chicago Area Transportation Study (CATS) in 1958 and presided over the introduction of a new level of complexity. CATS did not rely on drawing-board dreams of future land use but developed a model of the Chicago economy's likely development and future total and locational requirements for land. With advantage of hindsight this appears crude, but at the time it represented a major departure. The CATS model consisted of a function relating the percentage of population capacity of concentric zones to distance from the centre of the city in 1980. The CATS analysts, examining historical data, saw the urban growth process essentially as a great expansion of densities in an annular pattern around an inner city zone in which densities tended to decline; this trend was extrapolated and a density function for 1980 was produced. Along with current population and land use and the target year population estimate, this density function provided a distribution of households throughout the urban area in 1980. These projections of future land use then were fed into the traffic model, which predicted future circulation requirements.

As a model the CATS approach tells little about how the distributive processes shaping the urban pattern actually work. Its importance lies in its statement in mathematical language of a concept of how urban patterns change under conditions of growth.

The next big leap forward in transport studies was an heroic failure with a laudable objective. The Penn–Jersey (PJ) study, inaugurated in 1962, attempted to build a model to simulate the growth of the Philadelphia region, allowing for 'feedback' between transport developments, movement and land-use patterns. The traffic model, developed over the previous decade, was to be incorporated into a general model of regional change. The PJ study fully realised and stated that the target date, 1985, was just a convention and that a continual planning process was necessary, constantly updating projections of the future in the face of technological, unpredictable, and other *a priori* partially specified changes in tastes and capabilities. A simulation model was used, as it was impossible to understand causal relationships in a complex like a major metropolis in such a way as to calculate 1985 characteristics directly from 1962 knowledge. An attempt had to be made to construct a model in which significant processes by which a region changes are thought through and then expressed mathematically so that computers can be used to simulate human evaluation processes.

The spirit of PJ was willing, but its success demanded too much of technological capability and the willingness of the body politic. The PJ objective lay beyond immediate reach; aspirations exceeded abilities.

It seems that the technology of the land-use/transportation study has reached (it is hoped) a temporary plateau of elaboration in the 1970s and is now undergoing a period of consolidation while more attention is being paid to the political, social and economic questions involved in the specification of objectives for such exercises. At the same time, the theoretical underpinning of the methods and models employed has been enhanced.

The major components of the land-use/transportation framework as it stands and as it is applied today are as follows:

1. *Projections of Economic Activity and Population.* Attempts are made to estimate the future magnitude of activities in the study area. In some studies these have gone beyond the stage of simple demographic extrapolation and employ, at best, regional input-output models and, at least, economic-base studies.

2. *Land-use Forecasts.* The future population and activity estimated have to be located in the urban area. Models have been developed to simulate the process of urban development, drawing on location, urban growth, retail distribution, central place and interactance theories and optimisation methods.

The prediction of a future pattern of jobs, residences and services provides the input to the following elements which together comprise what is known as the 'traffic model' and include:

3. *Trip-end Estimates.* Forecasts are made of the number of journeys the various land users will generate as the beginning and end points of trips.
4. *Traffic Distribution Forecasts.* Having estimated trips originating and terminating as a function of land use, it is necessary to join the trip-ends as a distribution of traffic, an origin/destination matrix of flows.
5. *Modal Split Forecasts.* The flows established with a distribution model have to be allocated between the different transport media available.
6. *Traffic Assignment.* The flows by various modes on complex networks have to be assigned to specific routes.
7. *Design and Evaluation.* The objective of the whole exercise is to decide what is the best transport system to meet future requirements.

All these elements have to be considered as integrated components of a whole. The flow diagram, Figure 2.1, indicates their interrelationships.

Often, however, the models are given an air of individuality, as though the results of each constitute an end product. From a behavioural point of view it is meaningless to separate decisions to travel from the choice of destination; nor are decisions about residential location and the length, number and mode of trips arrived at individually. But, so far, it has been necessary to advance using successive approximations, feeding back results from one model to another in the hope that the system will converge towards some mutally compatible set of estimates. The theoretical searches which have to be made are for more behaviourally realistic models, which are more highly articulated with respect to the whole range of interrelated decisions which urban living demands.

There is an unfortunate tendency in transport studies, in the UK certainly, to concentrate attention on the traffic model parts of the process, accepting independent projections or plans for land use, employment and population. These predictions of future economic activity, population and land use are made independently of projections of future demand for movement and supply of movement facilities. The cost of movement in any urban area, relative to that in other cities, is likely to affect the rate of growth of its activities and population. The rate of growth and well-being of a community are not independent of the level of investment in its transport system. The way in which

Figure 2.1

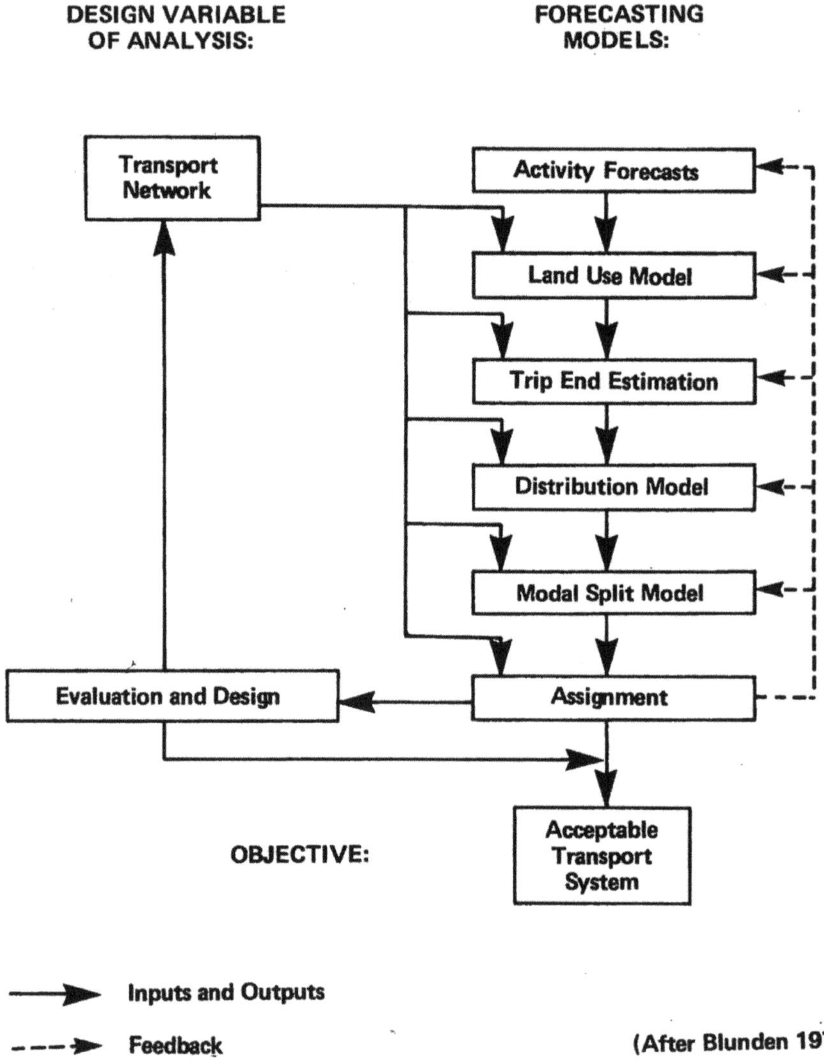


(After Blunden 1971)

activities are disposed geographically in a city is a direct function of the nature and configuration of its transport system.

We shall now take each element of the framework in turn and examine some forecasting methods used in more detail.

### *Economic Activity and Population Forecasts*

There is a perverse division in theory and practice between demographic and economic forecasting. Taking a broad historical perspective it is obvious that the growth of population is closely related to technological advance and the amount of wealth available for consumption and investment in societies. In the short run changes in sexual habits, medical technology and wars and disasters – and thus in the structure of the population – can have considerable effects on the demands placed on the economy. There is really no adequate theory specifying these relationships. Demographers have traditionally been satisfied with observing statistical regularities and extrapolating from mathematical expressions of these regularities. Work in economic history on the development process tends to be literary and offers little in the way of quantitative operational devices to assist in forecasting. The body of economic theory dealing with growth consists of tractable mathematical formalisms too abstracted from reality to offer any hints on the process of population growth. The operational models used in making economic forecasts usually depend on an exogenous projection of demand and thus of populations. And yet population forecasts should take some account of economic trends in society. At the national level the rate of migration, at least, will be affected by the comparative level of activity in an economy. There is evidence which suggests that the birth rate is similarly affected. Within national boundaries, where geographical mobility is usually higher, the effect of disparities in levels of economic activity on migration rates will be more marked. The best that can be achieved at present is an iterative forecasting procedure as in Figure 2.2, in which economic and demographic forecasts are coupled and made sequentially until some balanced and realistic picture emerges. Starting with a guess at a future population, estimates of future consumption requirements could be made, which could be translated into production requirements from which employment in various sectors of the economy could be estimated. Employment figures could then be used to modify the population forecast. The revised population could be fed back into the economic forecasting mechanism and so on until some equilibrium emerged.

Demographic forecasting consists either of curve fitting and extrapolation or the use of the components method in which cohort curvival rates are estimated and used to age the population through a projection period.

The impact of changing community needs on the economy can be traced with a scaled down version of the macro-economic theory of

Figure 2.2

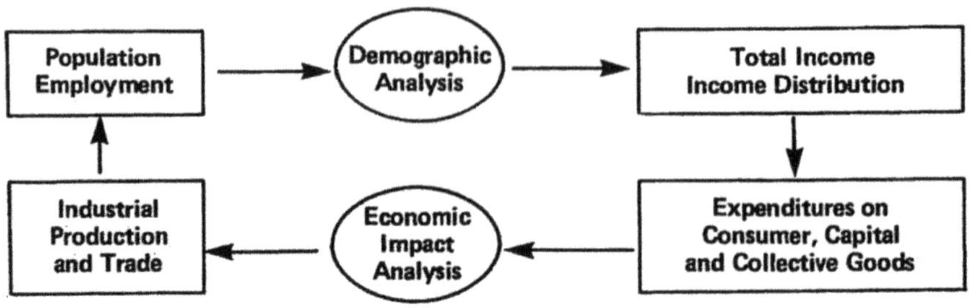


income and trade. The Keynesian multiplier, which finds the change in income (Y) resulting from an exogenous shift in demand ($\bar{D}$) in terms of the marginal propensity to consume local goods (s) with:

$$dY = \frac{1}{1 - s}\, d\bar{D}$$

has been applied to regional economics. However, since income data is seldom available for small areas, this model can only be implemented empirically by using employment data as a proxy for production and income in a region. Employment is divided into activities producing goods for export, designated 'basic' employment, and those producing for local markets, 'non-basic' employment. In practice this is accomplished by quite arbitrary assignment. Many sectors for which employment is reported obviously serve both local and external markets (considered retailing in a major metropolis) and there is no adequate way of making the basic, non-basic distinction. In addition, the ratio is also dependent on the size of the geographical units involved. The larger the region, the more internally cohesive the local economy and the lower the degree of interaction with other areas. Leaving these measurement problems aside, the proportion of income spent locally is assumed to be given by non-basic employment divided by total employment. If it is assumed that the marginal propensity to spend locally will not change with different levels of total regional income, then the multiplier can be expressed as the ratio of total to basic employment:

$$\frac{1}{1-s} = \frac{1}{1-(\text{non-basic employment})/(\text{total employment})}$$

$$= \frac{\text{total employment}}{\text{basic employment}}$$

This economic base multiplier has been used to forecast the impact of an exogenously determined change in the basic sector. The change is multiplied to give the future total employment. This does presume that exports are basic to the growth of a region, which is by no means a proven fact. The method is based on the short-run theory of income and cannot deal with changes in technology, wages, prices or resource availability.

A major objection to this model is the aggregate level at which it operates, ignoring the effect of changes in the *composition* of consumption. According to the theory, the income of a region will remain unchanged as long as aggregate consumption and trade remain unchanged. A change in their internal composition will not affect income. If productivity varies between sectors, then this will not necessarily hold. To account for the impact of such shifts it is necessary to decompose expenditure and production into finer categories. This can be achieved by making some simplifying assumptions about production functions and structuring the economy as a set of simultaneous linear equations in input-output analysis. With production relationships expressed as a matrix of input-output coefficients (A), a vector of gross outputs (X) can be estimated from any vector of final demands (D) with:

$$X = (1 - A)^{-1} D$$

taking into account both direct and inter-industrial effects.

In addition to the shortcomings of the multiplier and input-output methods when applied at the national level, there are some additional pitfalls involved in their local application. The finer the geographic scale, the lesser is the degree of isolation of the area in hand and the more important to the workings of its economy are trading relationships which lie outside the ken of Keynes' aggregates and Leontief's technical matrix. Input-output analysis works best for a large economy which is as closed as possible, i.e., with little overseas trade. Another difficulty is that the assumptions of the linearity of production functions and the temporal stability of technical coefficients over reasonable periods of time, which might be acceptable at the national level, are not so tenable at the regional level where the effects of increasing or decreasing returns to scale and the building of say one new modern plant may make themselves strongly felt in certain sectors, if not in many.

The CATS economic activity model (CATS, 1956) provides an example of how these methods have been used in practice. The CATS model was a four-stage sequence starting with given population to which were applied exogenously determined rates of population growth and productivity increase which determined regional income and income distribution. Income increases resulted in changed consumption patterns which are reflected in production by industry. The differential effects of consumption shifts are traced through the network of inter-industry relationships using input-output analysis to get estimates of outputs by Chicago industry, imports and exports from the rest of the US and the world. From output levels, employment in various sectors can be estimated and compared with population estimates.

*Land-use Forecasts*

Much land-use forecasting elaborates on the work of Lowry (1964). The Lowry model works in terms of the relationships between three basic types of land-use – places of employment, residence, and service activities. Accepting the location of basic employment as a given starting point, the model distributes workers from places of work to residential zones and converts numbers of workers in zones to total residential population. The combined distribution of employment and residents generates a distribution of service activities and thus more workers. These service workers are distributed around service centres as residents, who give rise to further demands for services and so on. The process is traced through the system of relationships until a state of equilibrium is reached. The functional relationships between the three activities determining the distributions are expressed in terms of the frequency and length of trips between them – journeys to work and use services.

The location of basic employment is taken as the given starting point in most land-use models. The intra-urban location of activities, which might be regarded as basic, manufacturing, and office employment, is very much a policy issue and is often severely restricted in choice by planning regulations. There is some revealing theoretical work on the competition for urban land and locations by different activities. Alonso's (1964) 'bid-price' surface formulation is full of insight but is difficult to translate into operational forecasting models.

Weber's classic problem of manufacturing location theory (1909) is to find a minimum transport cost location on a transport cost surface given a set of sources of inputs and markets, the weight of materials per unit of product (which is unity in the case of the product itself and some fraction proportional to sales if there are several markets) and the transport rate for each good. An analytic solution cannot be obtained for this problem. There is, however, a numerical, iterative method of solution (Kuhn and Kuenne, 1962). Weber's formulation is only applicable at the macro-geographic scale, the complexities of inter-industry relationships, access to labour and competition for land at the intra-urban scale are more than could be handled in such a framework. Alonso (1967) has formulated the problem in vector analysis terms so that effects other than transport costs can be incorporated, but it is not an operable approach. In fact, the intra-urban location of much so-called basic activity is determined by specific transport and site requirements or the need to be close to existing activities to enjoy external economies. So we can consider the locations of any expansion of such activities as fairly obvious determinable or subject to planning prescription.

Lowry takes the location of basic employment as given and distributes workers from work places to residential zones. Wilson (1970) has shown that this can be achieved with an 'attraction constrained' entropy maximising model of the form:

$$T_{ij} = B_j V_i^{\alpha} E_j e^{-\beta c_{ij}}$$

where

$$B_j = [\sum_i V_i^{\alpha} e^{-\beta c_{ij}}]^{-1}$$

such that

$$R_i = u \sum_i T_{ij}$$

where

$R_i$ = residents in zone i
$u$ = labour participation rate
$T_{ij}$ = employees living in zone i and working in zone j
$V_i$ = attractiveness of zone i for residential use
$E_j$ = employment in zone j

$c_{ij}$ = cost of travel from zone i to zone j and
$\alpha, \beta$ = empirically determined parameters.

Such a residential model may be disaggregated to take account of different segments of the population whose location behaviour might be quite different. In the entropy maximising procedure, the constraints imposed represent what is known about the behaviour of individuals in the system. Wilson suggests that for residential forecasting we need a family of expressions constrained in different ways to accord with different types of behaviour. He develops a hybrid model to encompass four kinds of location behaviour: (i) locationally-constrained people, such as immigrants; (ii) workers constrained to fixed residences (production constrained); (iii) workers constrained by fixed jobs (attraction constrained), as in the Lowry model; and (iv) workers constrained by fixed jobs and fixed residences (attraction and production constrained).

By distinguishing the first three groups of potential movers from the non-movers of the fourth group, potential movers are estimated within the model. Preceding the residential allocation by a submodel estimating potential movers improves the behavioural content of the model as this corresponds more closely to the two-stage decision process in which factors *internal* to the household, such as stage in life-cycle, prompt the decision to change house or job and only then are *external* factors, such as accessibility and attractiveness of different areas, considered in deciding where to move.

Wilson goes further to remove the simplifying assumption of residential models to date: that incomes are the same for all, that all houses are of the same type and cost irrespective of location and at the same time takes account of different wage levels by location. By making type and cost of housing variable, the supply side of the residential location equation is made explicit. He then shows how a final disaggregation can be introduced to account for social class variables to ensure that appropriate people are allocated to appropriate jobs.

The second element of the Lowry scheme is the modelling of service location. This can be achieved with an attraction-constrained gravity model:

$$T_{ij} = B_j V_i^{\alpha} R_j e^{-\beta c_{ij}}$$

where

$$B_j = [\sum_i V_i^{\alpha} e^{-\beta c_{ij}}]^{-1}$$

and

$$V_i = Q_i + F_i w^{-1}$$

such that

$$E_i = v \sum_i T_{ij}$$

The new variables included in this are:

$T_{ij}$ = residents living in zone j using services in zone i
$V_i$ = attractiveness of zone i
$F_i$ = available floor space in zone i
$w$ = minimum space standard per service employee
$c_{ij}$ = cost of travel between zone i and zone j
$E_j$ = service employment in zone i
$v$ = service employment to population ratio and
$\alpha, \beta$ = parameters to be estimated.

This model can be disaggregated according to a hierarchy of scale into neighbourhood, local and metropolitan levels and according to different types of services.

A wider range of styles in land-use modelling is available. The Herbert-Stevens model (1960) starts with a given distribution of industry and households at a base year and allocates increments to the household population by type to land available for residential use in a set of zones. The number of additional households and land availability are estimated exogenously for the forecast period. To carry out this allocation, exogenous estimates are required for each household type of residential preferences in terms of space, travel patterns and the budget available for residential purposes and also of the costs of providing each bundle of residential goods. The output of the model is the optimal allocation of households by type to zones of the city and the site rent prevailing in each zone for the forecast year. The model assumes that household type groups compete for available sites and that those who can offer the highest price for the land per unit area acquire it. This is formulated as a linear programme, the maximand being the total amount of income that can be spent on residential sites. Formally this is written as:

$$\text{MAXIMISE } V = \sum_k \sum_i R_{ki}(b_{ki} - c_{ki})$$

such that

$$\sum_k S_{ki} \leqslant L_i$$

$$\sum_i R_{ki} = N_k$$

$$R_{ki} \geqslant 0, \text{ for all } k, i$$

where

| | | |
|---|---|---|
| $b_{ki}$ | = | residential budget of household type k in area i |
| $c_{ki}$ | = | cost of residential goods and services, excluding site rent, for household type k in area i |
| $S_{ki}$ | = | space consumed by household type k if located in area i |
| $L_i$ | = | amount of land available for residential use in i |
| $N_k$ | = | number of type k households to be located, and |
| $R_{ki}$ | = | number of type k households located in i |

The $R_{ki}$ are the solution variables. The constraints require that the land consumed in each area not exceed the amount available and that all households by located. The dual of this problem is:

$$\text{MINIMISE } Z = \sum_i r_i L_i - \sum_k v_k N_k$$

subject to

$$S_{ki}\, r_i - v_i \geqslant b_{ki} - c_{ki}$$

$$r_i \geqslant 0, \text{ for all } i$$

In this minimand, $r_i$ is the rent per unit of land in area i and the $v_i$ are 'subsidy' variables which accommodate the possibility of numbers from two household type groups being allocated to one area. If this happens, the solution will assign the area the rent of the group with the highest ability to pay. The difference between the rent paying abilities of the two household types is made up by these 'slack' or subsidy variables.

The EMPIRIC model, developed for the Boston Regional Planning Project (Hill, Brand and Hansen, 1966), is essentially an extrapolation of current trends in order to allocate given total growth in categories of population and employment to zones of the urban area. The change in one of these categories of 'located' variables is held to be linearly dependent on changes in other population and employment variables and also on several 'locator' variables and changes in these. Thus, the model is a series of linear equations having the general form:

$$\Delta R_g = \sum_h a_{gh} \Delta R_h + \sum_k b_{gk} (Z_k \text{ or } \Delta Z_k)$$

where

$\Delta R_g$ = the change in the gth located variable for one time period
$\Delta R_h$ = change in all other located variables ($h \neq g$)
$Z_k$ = level of the kth locator variable
$\Delta Z_k$ = change in the kth locator variable in the time period of the forecast and
$a_{gh}, b_{gk}$ = coefficients to be estimated

For estimating $a_{gh}$ and $b_{gk}$, the model requires zonal data for at least two time periods on both locator and located variables. The five located variables, which appear on both sides of the equations, and which are allocated by the model, are the white collar population, the blue collar population, retail and wholesale employment, and a category of all other employment. The locator variables are measures of density, land use, accessibility by car, accessibility by public transport and the quality of water and sewage facilities of zones.

The San Francisco model (Arthur D. Little, Inc., 1966) is based on the micro-economics of the housing market and the premise that the suppliers of real estate respond to demand so as to maximise the yield on their investments. The model moves forward to the forecast date by iterating the market and other process over two year intervals. The forecasts produced are zonal estimates of overcrowding and excess demand by household type, the racial composition of zones, the distribution of substandard housing and of building and rehabilitation activity and of costs, expenditures, rents and profits. Exogenous estimates of household numbers differentiated according to size, race, income, occupation and rent paying ability are used to forecast the changes in housing stock from an initial stock differentiated by type, condition, and location. Each household type is characterised by a ranked listing of resident type preferences lying within their budget. These preference lists are matched with stock of available housing, and residences are assigned to households that can offer the highest price. If there is an excess of demand over supply, the price goes up and some households are assigned to their next best residence type choice. The price rise increases the profitability of that type of housing, which induces the building of more houses of this type or the rehabilitation of some stock in the next rank of the model. Thus, the quality and quantity of the housing stock declines or increases in response to the market prices. Market operations in the model may be modified by government intervention and so different policy measures can be evaluated with it. The whole thing is in essence a bookkeeping exercise. The preference lists of household types are established from preferences revealed in the houses different groups currently occupy. These preferences are extrapolated by assuming that a household would ultimately like to occupy a house type now occupied by a similar household in a higher income group. It is assumed in this model that accessibility to jobs or services does not influence the location decisions of households.

In addition to the change in the housing stock resulting from market and political pressures, the model also allows for the ageing of property at either a 'normal' rate, representing deterioration given routine maintenance, or at an 'accelerated' rate, under circumstances where the owner profits by not maintaining the structure. This ageing process is considered to consist of five stages: (i) sound; (ii) minor repairs; (iii) deteriorating; (iv) dilapidated; and (v) an absorbing state for demolition or removal from the market. Matrices with elements denoting the probability of transition from one state or the next over time were constructed for different housing types under both normal and accelerated ageing assumptions. The elements of these transition matrices indicate the probability of a structure staying in the same state or deteriorating by one condition from one time period to the next, thus non-zero elements may only appear along the main diagonal of the matrix and the diagonal immediately above it. The initial stock of housing is aged by applying these transition matrices. Given the structure of these matrices, an initial stock of houses would progress through the transient states of deterioration and eventually disappear in the absorbing state of demolition. However, the market and the public sector parts of the model continue to introduce new housing between successive applications of the transition matrices, so that the ultimate position of everything being absorbed in the fifth state is never reached.

*Trip-end Estimation*

Some of the above land-use models are couched in terms of trips and so the models themselves provide estimates of trip-ends. More usually, traffic analysts have to work from land-use plans and projections of population characteristics and non-residential activities for small areas, both of which are generally developed outside of, and often independent of, the transport-study context.

The goal of trip-end analysis is the provision of quantitiative estimates of trips starting and finishing in the various zones of the study area. People make decisions on how many journeys to make for what purposes in the light of their wealth, the means of transport available to them, their location with respect to other activities, and so on. It is unreliable to forecast the numbers of trips starting and ending in different areas directly, but the variables which determine them, income level, car ownership and location, are more easily predicted. Thus, if functional relationships, reflecting causal linkages, between the volume of trip-ends and the land-use and socio-economic characteristics of the population of zones can be established, then reliable forecasts of traffic generation can be made from land-use and demographic forecasts.

The three chief methods in use for relating generation to the relevant characteristics of analysis units of a transport study, be they zones or individual households, are: the establishment of generation rates for different land uses, the use of least squares regression analysis, and cross classification or category analysis.

The first method simply calculates trip generation rates for different types of land use, usually residential, industrial and service uses. Current land-use data are related to data collected in an origin/destination survey in terms of trips per unit area.

Regression is used to establish functional relationships between the number of trips per household or zone and a number of variables expressing household or zonal characteristics. The degree of association between the dependent and independent variables, measures of significance of the estimated parameters and of expected error, can be calculated to assist judgement of the validity and utility of equations for forecasting. The regression model is based on assumptions of identifiable, regular relationships and normality of distribution of values of the variables, which can be hard to meet.

In cross clarification or category analysis the change in one variable (trips) can be measured against changes in two or more other variables (say household income, car ownership, occupation of head of household, etc.). In essence, n independent variables are divided into several categories, creating an n dimensional matrix. Observations on the dependent variable (trips) are assigned to the cells of this matrix according to the categories of the independent variables with which they accord. The means of the values of the dependent variable observations assigned to each cell are then taken and these means constitute the response surface of trips over the independent variables. As Winsten (1968) has pointed out, this is a very proper form of regression analysis when there are enough data to avoid relying on least squares curve fitting and the necessary assumption of relationships which can be transformed to linearity. The technique has the desirable properties that the effect of one independent variable can be examined at various levels of the others and that qualitative variables, e.g. occupation, class, race or type of house, may be included in the analysis. For forecasting, household travel patterns have to be translated into zonal estimates of trip-ends, for it would not be computationally feasible for a realistic problem to load trips onto a transport network household by household. Predictions of numbers of households in various categories of the independent variables in each zone are applied to the base year trip rate matrix and the results are summed to give total trip-end estimates.

There has been a move in traffic generation analysis away from working in terms of the characteristics of areal units (zones) and toward the development of behavioural models of decision-making units – households or firms. There is a strong feeling that aggregate travel behaviour patterns might best be derived through the aggregation of the behaviour patterns of a large sample of individuals rather than by attempting to predict aggregate behaviour in terms of its relationship to other aggregate (e.g., zonal) measures. On the theoretical side, there have been efforts to produce economic models of transport demand but most workers have had to acknowledge the difficulty of sustaining the usual assumptions regarding markets and demand. The fundamental assumption of demand analysis, that units of output in relationship between price and quantity are homogeneous, is untenable when considering travel demand. The facts that car drivers have a joint role as producers and consumers, that transport networks are used jointly for a wide range of activities, and that the output obtainable by one user of a transport system is largely dependent on the behaviour of other users, militate against the development of economic models. Lack of knowledge of the interactions of people and goods, facilities and the location of activities has limited the development of viable operational, descriptive models of trip demand in conventional economic terms.

The technique of cross classification analysis, as used in trip-end estimation, is based on household behaviour. In regressional analysis households or zones have been used as observations. Stowers and Kanwit (1966) point out that in more conventional zonal unit equations what is being measured and explained in the model is the variance between zones not between households, the basic units responsible for travel behaviour. The between-zone variance is usually small in relation to between-household variance as observations on zones are mean values for all households in the zone. They note that since data are collected on a household basis much information is lost by areal aggregation at as early a stage as trip generation analysis. Areal aggregation can obscure much of the difference in social characteristics of households and, therefore, obscure relationships which may exist between these variables and travel behaviour. Further, relationships developed from aggregated data are sensitive to the size and degree of internal homogeneity of zones, which explains why regression results based on zonal units differ considerably between towns. Stowers and Kanwit conclude that analysis should preceed aggregation and the invariance of relationships with regard to zone size should be established before the analysis of zonal data is used for forecasting.

In generation analysis it is usual to stratify the dependent variable by trip purpose, according to the needs of the distribution stage of the traffic model. These needs are governed by the scope and objectives of the study and the size of the study area and nature of its transport system. Trips can be classed according to whether they emanate from the home to go to work, shop, school or to seek recreation, or do not start at home, whether they are commercial vehicle movements or taxi movements. A further stratification on the basis of mode of travel is possible. In large cities where public transport is widely used, it is usually to forecast 'person trips by purpose' and the estimate the proportion of future public transport users with a modal-split model either before or after the distribution phase. In smaller places, trip-end estimation may be carried out separately for public transport and car users or on car users alone if there is no public transport.

### *Trip Distribution*

There is disagreement about the ordering of the stages following trip-end estimation; as intimated in the previous section, some would put the modal-split phase before traffic distribution. In operation, traffic analysts usually allow for feedback between the models, however coupled, so that the results of a subsequent analysis flow back into a previous one and thereby achieve a satisfactory balance by a process of iteration. Here we will follow the order established in Figure 2.1.

There are two chief ways of forecasting the flows of trips between origins and destinations, of joining the trip-ends established in the previous stage. First, there are growth-factor procedures which expand a known travel pattern to some future year by growth factors reflecting land-use changes in zones of origin and of destination. The second method involves the use of distribution models which simulate travel by relating trips to the distribution of activities and the characteristics of the transport system. These models require calibration from current patterns of behaviour before they can be used for forecasting.

The chief among the methods used in distribution is the gravity model. Again, Wilson (1967) has derived this as the most likely configuration of flows between zones with known numbers of trips starting and finishing in them. His doubly constrained distribution model has:

$$T_{ij} = A_i B_j 0_i D_j e^{-\beta c_{ij}}$$

where

$$A_i = |\sum_i B_j D_j e^{-\beta c_{ij}}|^{-1}$$

$$B_j = |\sum_i A_i 0_i e^{-\beta c_{ij}}|^{-1}$$

The balancing factors $A_i$ and $B_j$ are the solutions for the Lagrangean multipliers associated with constraints preventing more trips being drawn from origins than are available or being sent to destinations than they require.

The intervening opportunities model is based on the assumption that a trip will remain as short as possible, lengthening only as it fails to find an acceptable destination nearer at hand. For a given category of trip, an equal area-wide probability of acceptance for any origin is defined for all destinations. All destinations or opportunities are considered in sequence according to their travel time from the zone of origin. In operation the first destination considered is the closest to the origin and has the defined area-wide probability of acceptance. The next destination has the same basic probability of acceptance, however, the actual probability is decreased by the fact that the trip has a chance of already having accepted the first opportunity. The process continues with each successive destination having, in effect, a decreased probability of being accepted. So, in this model, spatial separation is measured in terms of the number of intervening opportunities for trip satisfaction.

Distance friction is implicit rather than explicit as in the gravity formulation. The model takes the form:

$$T_{ij} = 0_i \, [e^{-LD} - e^{-L(D+D_j)}]$$

where

D = trip destinations considered before zone j
$D_j$ = trip destinations in zone j
L = measure of the probability that a random destination will satisfy the needs of a particular trip, this is an empirically derived function describing the rate of trip frequency decay with increasing trip destinations and increasing length of trip. The model is calibrated by varying the probability values until the simulated trip distribution matches the total travel time and percentage intrazonal trips of the base year data.

*Modal Split*

Modal split analysis attempts to forecast how many people will use the different means of transport available for trips between specific origins and destinations and for specific purposes. Modal split has been done in the course of forecasting trip-ends, estimating public transport trips originating and terminating separately from car trips. The market for public transport was seen as a 'captive clientele' defined in terms of car ownership, density of development, land use, age structure, and income levels of zones. A different view saw the process as more competitive and divided traffic between modes according to cost and service quality variables, after having distributed it between zones. The surplus of trips over some specified minimum volumes for car and public transport use was split on the basis of relative travel time or cost by the modes in question. It would be appropriate to combine both approaches, using the characteristics of the consumers and of the transport system to determine volumes on different modes. However, one effect which has proven difficult to handle is the influence of congestion which might arise on routes in the subsequent assignment procedure on choice of transport mode. One policy question which is frequently raised is whether car trips on heavily congested parts of the network could not be diverted to public transport. The answer to this calls for the development of direct feedback from assignment to modal split analysis.

The competition between public transport facilities and cars is a major debating point in urban affairs and it is not surprising that much research has been done to develop models to explain choice in terms of travel time, cost and convenience, and the nature of the traveller and his purpose, so that the results of considerable changes in the transport system can be predicted. Warner (1962) used regression and discriminant analyses to estimate the probability of choosing one mode over another as a function of travel time, cost and characteristics of the journey and person. In Britain, Warner's work was followed up by Quarmby (1967b), who expressed the problem in terms of the disutility of other means of travel. The traveller can be expected to choose the mode which minimises disutility. Discriminant analysis is used to find the best values of the parameters to explain the observed choices of a sample of travellers. Quandt and Baumol (1966) deal with both distribution and modal split forecasting in the 'abstract mode' transport demand model. Modes are characterised in terms of speed, frequency, comfort and cost, with no reference to the equipment or administrative structure of the service. A traveller's decision to travel and choice of mode is considered to depend on the absolute performance level of the 'best' mode on each criterion and the performance level of each mode on each criterion relative to the best mode. Since the model is couched in terms of 'abstract' criteria, the performance characteristics of new or proposed modes are merely inserted into the existing formula to forecast their use.

### *Traffic Assignment*

The trips distributed between origins and destinations and allocated to different modes of travel have to be assigned to specific routes of a transport system which is to be evaluated. The trips are loaded onto the relevant network at nodes located near the centre of each zone and their paths to destinations are traced through the links of the network, the number of trips using each link being accumulated as the assignment proceeds so that finally a density of traffic flow is produced. Assignment is the most powerful device of transport studies for testing designs for the improvement of the means of circulation and for suggesting where improvements are needed. It can be used to forecast traffic volumes on the existing network at some future time and to forecast the effect of improvements in the network so that the operational efficiency of the present system or proposed systems can be evaluated.

To start with, minimum travel time or minimum travel cost paths between zone centres are traced, building up a 'minimum path tree' connecting each zone with every other. The network is then loaded with the trips following their minimum paths to their destinations. Either trips between origins and destinations may be assigned to the one minimum path in an 'all or nothing' assignment, or trip volumes may be divided between the shortest route and the next best routes in a 'diversion assignment'. As the network is loaded, traffic on some links may exceed their design capacity, the resulting congestion necessitating a revision of the time values on links used in establishing minimum paths. If this occurs, the assignment may be interrupted for the re-calculation of minimum path trees before the assignment continues. This is known as 'capacity restraint' assignment. Capacity restraints make it necessary to decide which traffic is re-assigned and which remains on the original minimum path and the decision is incorporated in the procedure as the order in which flows between zones are loaded onto the network.

*Design and Evaluation*

The ultimate objective of the predictive efforts outlined above is design and evaluation. Forecasts are made to support the identification of that transport system which is optimal in some objective, quantifiable sense of performance. Design and evaluation are dealt with together in order to stress the duality of their relationship. What constitutes a 'good' design can only be known if there are reliable methods of measuring performance.

The evaluation of performance has traditionally been considered the engineers' province, wherein designs are tested by the assignment of traffic to the proposed system, with the objective of finding a transport system which will enable the given traffic to circulate most efficiently. As noted above, the assignment procedure indicates congested links on the network. Once these are identified, there are several engineering design options. It might be quite sensible to do nothing and allow congestion to operate, re-routing traffic and spreading congestion more evenly through the network. Increased capacity of new alignments might be introduced to improve efficiency, or controls and inducements to increase public transport usage might be tried, reducing vehicular traffic volume. The options are evaluated in terms of travel speeds, costs and service levels for the links in the network. Each proposal can be submitted to the assignment model and the total time or cost for all road users calculated. The approach is that of a traffic engineer and does not assist in answering the more important and far-reaching economic and social questions about the balancing of costs of system improvement and the benefits which accrue.

**Cost Benefit Analyses**

The objective of any investment appraisal is to maximise the difference between the present values of future gross benefit streams and the present value of future cost streams, both discounted at an appropriate rate of interest. The golden rule for investment decision making is to calculate the present discounted net value resulting from each possible decision and then to choose the one giving rise to the largest. In the public sector, where no market mechanism provides the articulation of utilities as prices, it has been found necessary to generate sets of pseudo revenue and cost streams to make decisions, in what has come to be called 'cost-benefit analysis'. The analysis consists of identifying the costs and benefits which are relevant to the problem in hand, valuing these streams, deciding on an interest rate to discount them to present values and dèlimiting the legal and physical domain within which an economically rational solution can be found.

It is widely recognised that a new or improved route will not only serve those currently using it and divert some traffic from existing routes, but also generate new traffic. How can these effects be taken into account in making additions or improvements to the existing capacity and structure of a transport system? The task of evaluation is essentially a matter of measuring the increment to national welfare resulting from a large number of possible projects, taking into account the interactions between parts of the existing and proposed network and selecting those projects which maximise this increment. This can be formalised as a search for a building programme which maximises welfare subject to a fixed budget or as a schema for calculating a rate of return on projects to compare with or meet the requirements of a central budgeting authority. In the UK, for example, the Treasury requires that public works achieve a rate of return of over 10 per cent or produce a positive net worth when streams of costs and benefits are discounted at 10 per cent per annum over their lifetime. By investing in roads or rolling stock, or any capital goods, present consumption is foregone in order to achieve increased consumption in the future. The net yield of such investment, the excess of future consumption over present consumption, does not attract us to commit funds without an additional inducement, because we have a preference for present over future consumption. Interest rates are the prices which reconcile the productivity of capital with the preference for consuming now rather than later. A unit of capital investment is evaluated by transforming the stream of future benefits and costs it generates back to some present value. This involves discounting by the interest rate. A pound's worth of benefits to be enjoyed next year is not worth a pound now, even if we make an adjustment for inflation and are dealing in constant pounds. In general at interest rate i any principle grows at a rate of $(1 + i)^t$ where t is the number of years. Inverting this gives us the expression for present discounted value of a pound's worth of cost or benefit arising t years from now: $1/(1 + i)^t$. In making investment decisions among a large number of different uses of limited funds, the rule is to act so as to achieve the largest present discounted value, selecting projects with the largest net present values. For a series of prospective costs $k_t$ and benefits $b_t$ of an investment over its lifetime of n years, we can select project where:

$$\frac{b_1}{(1+i)^1} + \cdots + \frac{b_t}{(1+i)^t} + \cdots + \frac{b_n}{(1+i)^n} > \frac{k_1}{(1+i)^1} + \cdots + \frac{k_t}{(1+i)^t} + \cdots + \frac{k_n}{(1+i)^n}$$

If the construction cost K is considered to accrue at some present instant and if the operating and maintenance costs are netted out of the stream of benefits, this reduces to

$$\sum_{t=1}^{n} \frac{b'_t}{(1+i)^t} > K$$

where $b'_t$ are the yearly benefits less the operating and maintenance costs.

**Social Benefit**

In 1844, Dupuit proposed the use of consumer surplus as a measure of the social benefit of transport investment. Consumer surplus is the sum of the differences between what people are willing to pay for a good in the last resort and whatever the price to be paid in the market is. In the economist's fundamental diagram this quantity is given by the integral under the demand curve above the price line. When Marshall (1920) promoted Dupuit's notion as a measure of welfare, he qualified its applicability by pointing out that it was valid if money has constant marginal utility. In addition, the completeness of the measures does depend crucially on whether the consumer spends more or less on the item in question after its price changes than before. If the price of transport falls due to an investment and the consumer thereby spends less on transport, he has increased disposable income to apply to other goods. He will gain additional surplus from the purchase of these. If he does not increase his consumption of transport at all he will have the whole cost saving to spend on other things. If his demand is very elastic so that he increases his total expenditures on transport as its price falls, he will have to reduce his expenditure on other things and lose the surpluses involved. Marshall held that for the most part such effects would be very small. In the same vein, Hotelling (1938) and Hicks (1956) held that as long as the price change involved was not big enough to make additonal income available for significant changes in purchases of other goods, it was a reasonable measure of welfare. Hicks did, however, devise theoretically correct but unobservable measures of welfare. 'Compensating variation' is the change in income which would be necessary to make the consumer indifferent to the change under scrutiny. 'Equivalent variation' is the amount of money which would leave the consumer at the same level of satisfaction as the price change. The first measure is appropriate to answering compensation questions, whereas the second is the consistent index of welfare changes in all circumstances. It is held by some that the observable consumer surplus will not necessarily move in the same direction as these theoretically satisfactory values in indicating welfare changes or measuring the welfare associated with alternative courses of action.

The most rigorous condemnation of the general use of this measure came from Samuelson (1947), showing that most stringent conditions had to be met in order for the marginal utility of income to be constant and, thus, for the area under a demand curve to measure welfare. Income increases due to a decrease in the price of one good must either be absorbed wholly by the good in question and all cross elasticities of demand must be zero *or* all income elasticities must be unitary. An approximation to these conditions was not good enough. A major commentator on welfare theory, Little (1950), agreed and dismissed the measure as a 'toy'. The latest detractors, Chipman and Moore (1978) reinforce Samuelson's point and further argue that when there are many price changes, consumer surplus, being measured from a path through price and income space, can assume different values depending on the order in which changes in prices and incomes are traced.

The current proponent of consumer's surplus, Willig (1976), taking compensating and equivalent variation as the theoretically correct measures, has shown that the observable area under a demand curve is a good approximation. He derives upper and lower bounds on the errors involved which can be calculated with knowledge of consumer's base income and income elasticities of demand. He holds that in most circumstances the errors will be very small, overwhelmed by the errors involved in estimating the demand curve in the first place.

One matter which has not been fully explored is the relationship between individual and aggregate surplus. The theoretical formulations are usually phrased in terms of the individual. If the finding is against the magnitude at the individual level, clearly it is inoperable at the aggregate level. Findings in favour of the measure are driven to assume that consumers are similar so that the aggregate can be treated as one individual. The assumption of homogeneity of people is always cause for unease.

## Consumer Surplus in Practice

Whatever its theoretical shortcomings, the differences between reserve price and market price as a measure of the social worth of discrete changes in economic circumstances is widely employed in the appraisal of transport investments and policies. Indeed, the idea first saw the light of day in the *Annales des ponts et chausées.* In choosing between transport investment and policy options, the difference between the consumer surplus accruing when nothing is done and after a change is instituted has been taken to constitute a measure of the benefit of that change. This approach has the added attraction that it can be reasonably estimated from two observations on price and quantity pairs. If the demand for travel increases as price decreases from p to p′ then the consumer surplus associated with the improvement is in general:

$$\Delta CS = \int_{p}^{p'} T(p)\, dp$$

Road space, however, is a non-marketable, essentially collective good. There is no market price which registers what use of a piece of road is worth to the marginal customer. There are no observations on price and quantity demanded from which to estimate the shape of a demand curve. Despite the absence of a market, road and public transport users do vary their use in response to a 'price'. The price paid includes expenditure of the traveller's time, including that spent in walking to and from parked vehicles. When we consider public transport, waiting time is obviously relevant. The price also has a cash component including tolls, parking charges, fares and vehicle running costs. This combination of time and money outlays is referred to as the 'generalised cost' of travel. In making choices between means of travel or routes or indeed whether to travel at all, the consumer behaves according to his perception of these costs. Observations of users' choices between routes and modes with different time and cash costs will reveal the value placed on time and how that varies between different uses of time. In the UK, it has been found that, in general, people value non-working time spent in a vehicle travelling at about 40 per cent of their wage rate, while time spent walking and waiting is twice as costly. If we wish to predict demand or choices then clearly these behavioural costs are the appropriate explanatory variables.

However, economic evaluation of a policy or investment is concerned with estimating the real resources consumed in travel and transport. The resources used up in travel may vary from those perceived by the consumer in two ways. The prices that people face may not reflect the resource cost involved. For example, in the price paid for fuel there is an element of taxation not representing a resource cost. Frequently the fare charged on public transport may not reflect the cost of operation either in total or in part. In addition, people's knowledge and perception is imperfect. People's behaviour suggests that the average estimate of the cost of running a car is somewhat less than a precise engineering assessment of the marginal mileage cost. These departures should obviously be adjusted for in making decisions on the allocation of resources.

If we leave these niceties aside and assume perfect knowledge and marginal cost prices, then the benefit of a transport improvement can be quite simply measured. An investment is undertaken to reduce the generalised cost of travel. If travel is a normal good this should result in an increase in demand. We can observe the generalised cost and demand for travel before the improvement, $c_1$ and $T_1$ in Figure 2.3. If we can predict these values after the improvement is in place, $c_2$ and $T_2$, then we can measure the change in consumer surplus over doing nothing to compare with the cost. If we assume that the demand curve is linear over the relevant range, then the integral $\int_{c_1}^{c_2} T(C)\,dc$, the shaded area in Figure 2.3, can be evaluated as:

$$\Delta CS = \tfrac{1}{2}(T_1 + T_2)(c_1 - c_2)$$

This measure of worth has become embedded in the evaluation of transport projects, where the public sector either controls or subsidises provision of the network. It is fused into the administrative procedures governing road and transport infrastructure appraisal in the UK (McIntosh and Quarmby, 1970) while the World Bank (van der Tak and

Figure 2.3

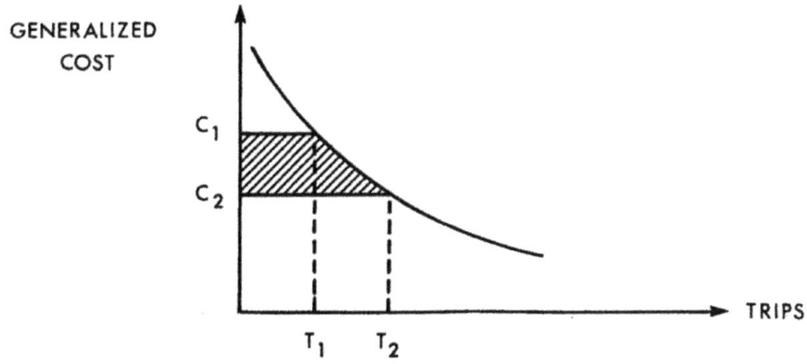


Ray, 1971) proselytise its use as a benefit measure for transport.

There are arguments about the operability of consumer surplus in a transport planning context occasioned by the network nature of the good. Harvey (1968) has objected that the measure takes no account of shifts in the demand curve for trips between an origin-destination pair, as trips to substitute or complementary destinations are generated or suppressed by the introduction of a new road scheme. Neuburger (1971) contradicts this with a demonstration that it is still the correct measure of benefits, even allowing for such shifts.

The assumptions about income elasticities and substitutability upon which the measure stands or falls are frequently concerned with choices between locations for plants or homes in the long run and forms and frequencies of travel and destinations in the short run. There are some impact studies in the wake of the Interstate system but most work concerned with urban areas is in the nature of forecasting on the basis of not wholly substantiated hypotheses. On the broader geographical scale, efforts to isolate the effects of increased accessibility on regional growth (Harris, 1974; Dodgson, 1974) have not provided convincing evidence of the power of this factor. Fogel (1964) conjectured that even an investment as large as the railroad system in the US, though it did result in large geographical transfers, did not have a major independent, embodied effect on national welfare. The same growth of wealth could have been achieved with a more compact economy geographically and a road and water transport system. The point which arises is that both the disposition of land uses and the transport costs expected in the absence of the decision in hand are data which should enter the calculation of welfare improvements. Increases in demand for transport can arise from a more extensive use of land to produce the same output, the benefit accruing entirely to land owners or transporters, if some degree of monopoly power exists in that sector.

If they are efficacious, the results of transport improvements are to be seen in both travel behaviour and land use, one being a reflection of the other. In a caution against double-counting benefits by including travel and other market effects, Harrison and Holtermann (1973) showed that the change in consumer surplus observed in the transport market merely summarises the surplus changes in all markets for goods and factors whose prices are influenced by the transport improvement.

They also pointed out that the ultimate incidence of the benefit may not be entirely upon those who travel on the facility. If the transport sector is monopolised it could reap the entire cost saving as profit and pass nothing on, giving rise to no increase in demand or rent. In classic rent theory, the benefit is gathered entirely by owners of land as rent. In reality much of the advantage may indeed pass on to other than the purchasers of transport. Among industrial users a firm with a heavy transport input may enjoy a drop in production costs. Depending on the responsiveness of demand for its product and the competition it faces, the firm may retain all of the saving as profit or pass it on to some of its customers in lower prices. Property owners will garner the benefit of increased accessibility if it causes property values to rise, even if they do not travel on the new facility. Transport benefits to renters may be reduced by the amount of any rent increase property owners can impose to gain some of the benefit. The benefit to home owners will accrue partly in the form of increased property value. In temporal terms the benefits fall to those in possession when the improvement is put in place, even if they sell, for the price will include the present value of the expected stream of transport benefits.

In geographical terms, aiming benefits at certain localities by the placing of a transport facility will not necessarily improve the lot of the occupants since the benefits may be passed on to owners of property, firms, and consumers at large. The geography of these transfers would seem of interest both academically and politically. It is generally accepted that welfare judgements must weight distributional as well as efficiency implications. There is in addition a legitimate political requirement for equity between territorial constituencies. The fact that the benefits of transport or other public works spill over jurisdictional boundaries and that exclusion of non-residents from their enjoyment is impossible, means that a given geographical disbursement of funds may not imply a similar disposition of benefits. For example, the improvement of rail commuting facilities within a city's territory may largely benefit suburban commuters. Even though the effects of losses and gains in relative accessibility may cancel out, so as to signify little

in terms of national or regional increases in well-being, they can be large changes to those immediately involved and can, thus, signify a great deal in local and state politics, as witnessed by the Crosstown Expressway controversy in Illinois. Aggregate consumer surplus is inadequate evidence to judge such cases. A map of the final incidence of benefits is the minimal supplement required.

Harberger (1971) justifies consumer surplus as available though imperfect, being far better than the commonly accepted national income measure which deals only in market prices. Consumer surplus at least incorporates second order effects of any change while national income represents only the first order effect. It does, however, rest on the premise that the benefit to each member of society can be added up without regard to the individuals who enjoy them. There are circumstances where this is within the bounds of political prudence, indeed a politically desirable stance. Despite recent departures, it remains an underpinning of representative government that undue discrimination between citizens is undesirable. A practical example of the application of this homogeneity assumption is the decision by the UK Minister of Transport voiced in 1969, to employ a single value of time for the calculation of transport benefits, implying a constant and common marginal utility of time and income. For universally provided facilities, say in funding local road improvements, where the investments are spread geographically pretty much as is population, and where the benefits are small and diffused to many people, an assumption of homogeneity and no regressive bias in the benefits seems reasonable. In deciding on the allocation of a fixed budget among a large number of similar uses, consumer surplus, which can be estimated from known present volumes and speeds and good guesses at future volumes and speeds, does seem likely to provide a consistent and unbiased ranking criterion. For large projects, arising more sporadically in time and space, whose effects are geographically confined, say public transport or motorway facilities, where there are potentially stronger locational effects, total consumer surplus may be inadequate as a welfare criterion. It will fail to signal distributional outcomes which may be politically crucial.

The theoretical employment of consumer surplus as an optimand to compare with a competitive equilibrium is subject to the same strictures. If the real circumstances being addressed do not represent too violent a departure from Samuelson's conditions, then consumer surplus may be a satisfactory measure of social desirability.

In several transport studies, cost-benefit calculations have been undertaken for the whole of a design or parts of it, but usually after the event as supporting evidence for the design chosen. CATS did incorporate a cost-benefit calculus in their planning process, comparing rates of return of different plans in terms of the ratio of travel-cost savings to investment costs. On the whole, the fact that precise specification of an objective function and the evaluation of costs and benefits is very revealing politically, and that it leads on to arguments about notions of 'welfare' and 'utility' frightens off the politicians in charge of transport studies from economic evaluation.

Both the traffic engineers' assignment and cost-benefit calculations are partial methods of evaluation, accepting a given disposition of activities and thus demands for interaction in the urban area. A more general evaluation would involve an objective wider than merely the transport system, seeking the optimum form of urban settlement. The interrelationships between the transport system and the disposition of activities would have to be specified and modelled. Land-use forecasts would have to respect the likely effects of different transport system designs and these different land use forecasts would provide a basis for judging between transport proposals. Despite valiant efforts, not much progress has been achieved in modelling their relationships. The approach throws the onus for ultimate assessment of plans and systems even more firmly at the feet of politicians and aesthetic experts, asking them to state clearly what 'desirable urbanity' consists of, that it might be optimised.

The network approach acknowledges the interdependence of pieces of road in an integrated whole. That the effects of improving a road are not confined to its users alone and are related to the structure in which it is embedded is particularly striking in urban areas. This realisation first elicited an engineer's response in planning practice. Proposed improvements were evaluated in terms of their contribution to overall network performance in carrying a given load. The key tool of this was the assignment algorithm which found an equilibrium between desires to move from point to point and limited link capacities subject to increasing congestion costs. As the feeling for the symbiosis of location and movement grew, so forecasts which tried to incorporate their reactions on one another more generally were attempted.

Having constructed descriptive models of transport demand and supply relationships with which to calculate the social welfare associated with proposed investments, it is not a far step to take a normative, prescriptive stance, seeking the network configuration which maximises welfare, however defined. With the flowering of mathematical programming methods, this became a viable proposition. We will return to this matter in Chapter 6.

**A National Systems Model for Transport Planning**

The urban transport planning model provided the inspiration for extending the systems model to the planning of regional and national networks (Kresge and Roberts, 1971). This involved predicting the developmental process to obtain estimates of the welfare change generated by alternative transport investments and policies. The procedure is outlined in Figure 2.4.

Figure 2.4

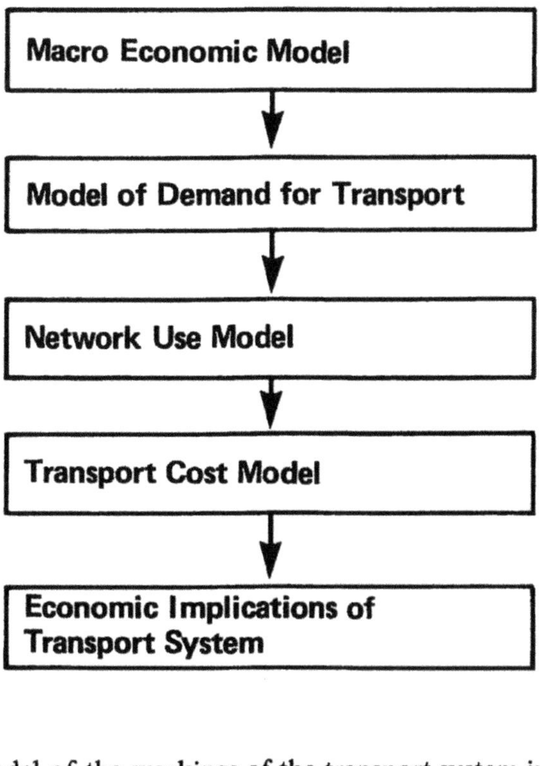


It places a model of the workings of the transport system in one of the workings of the national economy so that the ramifications of different transport investment and regulation strategies and projects can be traced through to their effect on gross domestic product and other measures of national progress and well-being. Final demand for various goods is established for estimates of consumption, investment, government purchases and exports and is then translated into native industrial production requirements and import demands by means of an input-output model. Demand and production are disaggregated on a regional basis and inter-regional trade in various goods is estimated with a gravity or programming formulation, according to the degree of homogeneity represented by each commodity class for which data are available. The transport system is represented as a comprehensive network of several modes of transport, including links for transfer from one mode to another. The cost characteristics of the links are determined and traffic between origins and destinations is assigned to various links in the network so as to minimise the total cost incurred. The costs incurred and the performance of the transport system can be transmitted back into the macro-economic model via the input-output coefficient matrix. Transport is regarded, along with households, government and the rest of the world, as an exogenous sector. There is a marginal row and column attached to the inter-industry coefficient matrix representing the inputs of transport to various sectors and the outputs that transport draws from various sectors. In this way the effect of investment or regulation, changing transport costs and their effect on various summary measures of the status of the national economy can be evaluated. Thus different transport investment strategies can be tested in terms of their overall effect on the economy. The model was developed and calibrated with data for Colombia and was used to examine transport investment impacts in that country. A series of transport investment plans was designated including no investment in transport facilities, concentration of funds on road investment, the building of one major new road link, and the concentration of funds on various rail projects. The road investment option proved superior to all others both in terms of benefits from reduced transport costs and also in terms of gains in gross domestic product.

**Partial Scheme Appraisal**

Notwithstanding these developments in the analysis of transport networks as systems, the economic appraisal of much rural, intercity and even some urban road capacity continues to be carried out under the implicit or explicit assumption of the independence of links in a network. These partial methods make an assumption of project independence and calculate the measure of welfare improvement for alternative solutions to a problem. The relationship of the investment with the rest of the economy registers through the discount rate on a cut-off value for the net worth of each scheme. The rubric attends to the margin, on the presumption that the latest additions proposed are the most marginal.

The most usual formula is the calculation of a first year rate of return, expressing user benefits in terms of changes in the cost of operating vehicles and reductions in delay and accident costs as a percentage of the construction cost of the scheme. In essence, traffic is extrapolated to the first year of operation with no allowance for generation, distribution or modal split differences between the base situation and the proposed improvement. Some reassignment between the project and its competitors is usually allowed for. The Department of the Environment (1972b) have elaborated this procedure under the title COBA. The calculations are extended through time to enable the calculation of a net present value discounted over the project's life. The calculation of effects on neighbouring parts of the network can be extended to 100 links. The influence of projected complementary or competitive schemes which result in reassignment can be fed in up to 10 years after the scheme's completion. COBA also offers improvements in terms of the more detailed treatment of junctions and the disaggregate specification of flows over time and by vehicle composition. The method obviously employs parameters, including speed-flow functions, relevant to rural conditions. Additionally, it is inappropriate for urban schemes inasmuch as it is a fixed trip matrix method with a potentially unsophisticated estimate of reassignment, taking no account of possible differences in total demand, trip-end distribution and mode selection between the base and the scheme.

Tindall (1974) devised a scheme evaluation technique for use in urban areas. The level of design detail required for its operation is, however, in excess of that available from the strategic, network design phase of planning which we address here. Also, the extrapolation of existing flows on an alignment which the procedure entails is quite inadequate in dealing with complex and widespread network effects.

## The Treatment of Uncertainty

Whether the network perspective is employed in its descriptive or normative form or whether projects are treated as independent, investment in roads is obviously made in the face of great uncertainty. The capital works contemplated are very long lived and fixed in place. History suggests that they may be vulnerable to radical technical change and potential obsolescence. The relations of concern between land use and accessibility are as yet poorly understood and not well quantified. Positive and negative effects of any investment in roads are non-marketed externalities of very uncertain worth, subject to changing collective values as well as changing personal preferences and auxiliary input costs. Many of these issues are matters of virtual ignorance, not mere uncertainty.

In practice, ignorance of the future has been dealt with according to the mood of the time. Expansionist optimism took leaps in the dark, catering to unknown potential by lavish overprovision. Dour, guarded frugality suffered the unexpected as its cost mounted in bottleneck overspills. The only analysis which can be brought to bear involves the extrapolation of current experiences or guessing what the future may hold. When degrees of uncertainty can be distinguished, degrees of arbitrary conservatism have been employed. Premiums were placed on discount rates according to riskiness which was perceived to compound with time. The life of the investment was foreshortened when it was feared it would fall prey to sudden obsolescence. Pessimistic benefit forecasts have been taken if the danger was seen to lie in the erosion of demand over time.

As the road investment process was formalised, so the import of ignorance about costs and demand for public attitudes became clearer. Quarmby (1967) seemed to be groping for a means of incorporating aversion to risk in cost-benefit calculations. He suggested that the cost of overestimation of benefits was higher than the cost of underestimating them, prescribing conservatism as the appropriate stance. The appropriate practice, he suggested, was to 'write down' benefits or set a higher cut-off rate of return, where this criterion was employed to allocate less than a fixed budget.

These *ad hoc* judgemental allowances, along with foreshortening the life attributed to assets and putting a premium on the discount rate, have been widely advocated and employed as practical adjustments for degrees of doubt. Writing down the values of costs and benefits year by year is considered appropriate where the risk of fluctuation is thought to be irregularly distributed through the future. The life span over which benefits and costs are discounted is frequently curtailed from an engineering estimate, if it is felt that the danger is of a sudden truncation of benefits. The use of a risk factor seems appropriate where doubt increases strictly with time.

Another practical tool, sensitivity analysis, was proposed for transport planning by Bergendahl (1969). He advocated the comparison of pessimistic and optimistic guesses as to the future environment of an investment. In general, the method involves varying assumptions about streams of benefits and costs and investigating how responsive present

Figure 3.1

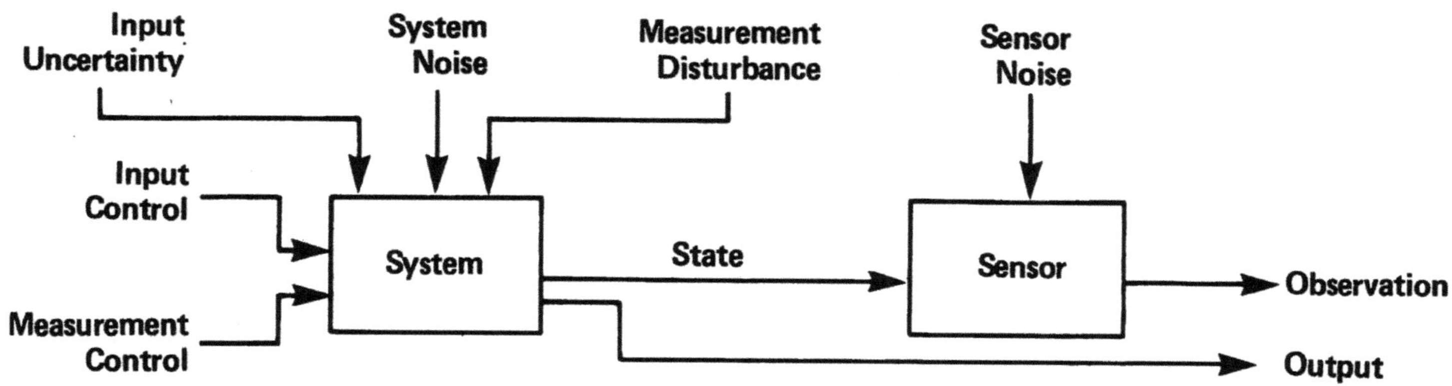

Input Uncertainty
System Noise
Measurement Disturbance
Sensor Noise
Input Control
Measurement Control
System
State
Sensor
Observation
Output

takes its toll within this structure.

Taking up the terms of the schema outlined in Figure 3.1, a *system* consists of a group of entities which are meaningfully connected by relations and activities upon their existence and attributes, which have a satisfactory collective boundary, and which interact for a common purpose. A *model* represents the system in a well-defined operational structure for the study of the common purpose. That is, the system consists of entities whose internal mechanisms are not relevant for purposes under investigation – traditionally labelled 'black boxes' – and which possess input-output relationships with each other. The system itself possesses an input-output relationship with the external world, its environment. Operationally, systems theory is a conceptual tool employed by model-builders to help design the 'best' system to achieve the purpose for which the system is conceived to exist.

Needless to say, the definitional use of 'meaningful' with respect to a system and the operational characterisation of 'satisfactory' in a model hinge upon the careful specification of 'purpose' and the critical analysis of the decision-maker's ability to influence control. In this we are treating modelling's use for 'problem-solving' rather than for 'problem-finding'. Of course, problems must be found before they can finally be solved, or at least they must be identified as problems before the solution can be deemed appropriate. However, it is the essence of meaningful planning that problems should be prevented from arising. Therefore, 'problem' as construed here must carry the broader concept of potentiality as well as actuality. We will return to this point shortly in the context of planning. Its importance here is that the systems are defined and models constructed for the study of some 'purpose' of consequence – it is this purpose which gives both endeavours meaning.

Measurement is the means of gaining knowledge about the performance of a system and the dynamic aspects of its internal interactions. This act of measurement is accomplished by a *sensor*, the sensor's output being termed an *observation*. The intent of the sensor is to ascertain the state of the system, where the state is the result of all past inputs to the system. Unfortunately, the internal workings of a system are not always directly observable and the observation must then be upon some set of variables, whose relationship to the actual state is considered known and is usually taken to have a quite restrictive functional form. The act of measurement disturbs the 'normal' behaviour of a system which introduces uncertainty into an observation on that behaviour. The accuracy of the observation is also disturbed by uncertainty in the specification of the input.

The performance of the system depends on and is evaluated with respect to the interaction of the system's *output* with the exogenous environment. The output is usually modelled as a known function of the state, although the output may not itself be a component of the observation. Therefore, to affect desirable performance, input *control* is manipulated (within certain physical and financial constraints) to achieve a desirable state and hence a desirable output. Since measurement is employed to monitor the behaviour of the system, control may be exercised through manipulating the sensor (again, within certain constraints) to control how the observations are made.

In the same way that the overall behaviour of a system may be analysed, the process of planning itself may be viewed within a system-theoretic framework. For decision-making considerations, the process of planning may be partitioned into modelling, analysis, estimation, and control.

*Modelling* is the most critical step in the process and is characterised by the development of a conceptual system (the model) corresponding to the real-world system. The model is constructed to gain an understanding of the behaviour of its real-world counterpart. The use of the model assists in selecting controls aimed at an optimisation of the quality of the real world. Therefore, if the model is not a good factual representation of the real-world system, any subsequent manipulations of the model are meaningless in the context of that system and any model output is potentially misleading in application.

*Analysis* is concerned with the behaviour of the system's output in both qualitative and quantitative aspects. The quantitative analysis determines 'precisely' how the output behaves, e.g., the functional relationship of output with time; the qualitative analysis determines general properties of this behaviour, e.g., its stability. Analysis is also concerned with the properties of any impinging disturbances, but only as they influence the system's state and not necessarily with the structure or source of the disturbances themselves.

*Estimation* is concerned with using the observations in calculating, deducing, inferring or guessing the properties of the real-world system. Generally, the observations are used to estimate the state of the system as a function of time under the assumption that the model is perfect. However, since it is impossible to construct an *a priori* completely-specified model for any future system, a set of observables is sought which represent the state of the system. The system can then be modelled by imposing some structure upon this set of observables. A system may be said to be *identified* or modelled when it has been completely specified in terms of the set of observables and the relations thereon, instead of the actual states and their relations; and the problem becomes the measurement or estimation of the observables. Unfortunately, it is not usually clear what constitutes the states of the system, and therefore, what observables to select as representatives.

*Control* is concerned with the selection of values for the control variables. From a general point of view, control variables are system parameters which the decision-maker can vary to achieve the desired obectives: e.g. a control input state in an allocation problem is the amount of the resource available and the decision is the amount to be allocated. The type of control problem which comes closest to reality in its potential for application in investment-making models is *adaptive control*. The adaptive control procedure is characterised by explicit incorporation into the model of some means available to the decision-maker for modifying the sequence of inputs in order to make the system more acceptable – casually viewed as 'changing one's mind'. Use of adaptive controls is desirable because the system's operation, the performance criteria, or the constraints may have been incompletely specified initially or may simply change. Also, as time progresses and more information is gained by monitoring the system's behaviour, the representation and estimation of uncertainty in the model require re-evaluation. Of course, any re-evaluation of the supporting tools of control determination falls back on modelling, analysis, and estimation.

Modelling is definitely the most important and critical aspect of a systems theory of planning, for, if the model fails to effectively represent the system, any subsequent analysis, estimation, or control determination may be misleading in application. The most important part of modelling is the specification of what problem is to be solved. The choice of model depends on what question is being asked. The object of the process is to find a model to aid in solving a problem, rather than to find a problem to experiment with a model. Analysis is the second most important area, for often the ability to effectively analyse the situation is all that is needed – especially when coupled with professional experience and judgement. The relative importance of estimation and control depends on the extent of uncertainty. For most situations involving only negligible uncertainty, control is obviously the more important; but as the degree of uncertainty increases, estimation and monitoring become more and more important to insure that everything is behaving as intended and to enable adaptation and re-evaluation in light of new knowledge. Faludi (1973) characterises the extremes on this continuum of importance: 'where the control variables are dominant, the subject is in command; where the constraints dominate, the subject drifts, waiting for the first clues from the environment, so he can form an image of the situation'.

**Uncertainty and Planning**

The analysis of the role played by uncertainty and its explicit confrontation in determining the choice of initial control is the dominant concern of this book. Estimation theory has traditionally been invoked as providing the needed tools for identification, adaptation, and decision-making problems. The planning framework advocated here attempts to formulate an explicit analysis of the influences of initial control determination under present uncertainty of future environments rather than circumventing such considerations by translating the problem into estimation of subjective probability distributions.

Control theory plays an overtly minor role in this exposition, although it is conceptually the *raison d'être* for the effort, for it must be assumed that the decision-maker's desire to control some system and the intent of the analysis of uncertainty is the judicious allocation of resources to effect control. The biggest, and sometimes effectively insurmountable problem, is to use available data to find out what the system is doing: to estimate its present state. If the system's state can be estimated to some reasonable degree of accuracy and its operational mechanism effectively modelled, then the desired control is often obvious – or can be determined by deterministic programming or control methods.

The position taken here is that the foremost element in control determination in the presence of uncertainty is analysis. Even programming approaches which purport explicitly to handle uncertainty begin with an extensive amount of deterministic information – the following being typical: 'Suppose a planner facing a *fixed* budget is considering expanding the capacity of a network containing a set of links and nodes whose individual capacities may be augmented at a *given* cost . . . In addition, a set of demands for a homogeneous commodity in the form of a probability distribution *exists* at various destination nodes.' (Midler, 1970, p. 267; emphasis added.) Then, what started out as a control problem – the investment sequence for network augmentation – becomes the difficult and critical problem of estimation or analysis of the uncertainties.

Deterministic programming approaches (increasingly the conventional practice in the planning of transportation systems) derive some deterministic growth in demand over the relevant region, construct the optimal network of the forecast terminal point value, and determine the immediate commitments toward this network within the deterministic environment. Within such a framework, the sensitivity of the 'optimal' network sequencing, if it is in fact discussed, is carried out with respect to the terminal state (e.g., Ahmed *et al.*, 1976). At this final stage, the total network performance is least sensitive to forecast errors because of discounting to present value and to the actual initial investment set because of the small number of standard road types, the lumpiness of investment, and, as a result of these two, the large link capacities within the network for significant periods within the planning horizon.

An extension of this purely deterministic approach is to compare the investment sequence derived from a 'best guess' about the future with sequences resulting from an 'optimistic guess' and from a 'pessimistic guess' (e.g., Bergendahl, 1969). This approach traditionally assumes stable, monotonic behaviour of the system with the decision to invest understandably relying on the 'best guess'. Currently, the mechanics of optimality assume some known behaviour of forecast variables and known relationship between them, even in attempts at probabilistic formulations of a specific investment project's behaviour (Cole, 1965; Pearman, 1976) and of general network behaviour (Charnes, Kirby, and Raike, 1966; Midler, 1970).

The complication is, of course, more fundamental that *what values* certain variables will assume in the future; it is the identification of *what variables* and their relationships and values most effectively represent the system. Perhaps, the most fundamental underlying complication of uncertainty is the inability to measure certain relevant elements of the socio-economic environment which influence the evaluation of the system's performance.

Based on the notion of measurability, a relevant variable may be classified as a *predictable* if it may be reliably measured and enough is known of its behaviour to forecast its future values or as a *goal* if enough resources (one of which is time) are available to effect its achievement. The arguments in support of the investment strategy espoused in this work are predicated upon this distinction between prediction and aspiration and, further, between the divergent nature of a prediction to reliably delineate the future and the convergent nature of an aspiration to control the evolution of that future.

A prediction, statistical or otherwise, is valid for detailed estimates of costs and benefits over a planning period that is limited by a desired level of confidence, probabilistic or otherwise. Under any reasonable level of confidence, this period cannot realistically be very long relative to the necessary horizons surrounding the evaluation of large-scale, long-lived investments within complex systems. That is, the range which may be taken to represent a given level of confidence under an *a priori* distribution must expand with time and, after some not too lengthy a period, this range becomes meaningless as a planning input. Alternatively, the level of confidence may be variable and associated with a fixed desirable range but, again, as time progresses the degree of confidence that the system will evolve within this range diminishes to meaninglessness.

On the other hand, an aspiration is a goal that lies a long way beyond the period of meaningful predictability and is based in part upon a broader notion of prediction of the constraining environment. An aspiration is characterised by qualitative and not quantitative statements about attainable and desirable states which are not necessarily inertial consequences of present system dynamics.

Under the traditional notions of justification for large-scale capital investments, initial commitments need to be governed by reliable forecasts. However, the actual measurement of such long-term optimality rests in an environment which is construed only as a set of goals at the time of these initial investments. That is, the traditional need for quantitative justification, if it is to be meaningful, must depend on forecasts of future values which are deemed reliable, but for the investment to be realised as the optimal, selection depends upon the decisions of future agents in achieving those identical goals that are set out in support of the initial investment.

It follows that pursuit of traditional notions of optimality under uncertainty is illusory. However, something must be done toward maximising effectiveness in the use of existing resources. The acceptance of modelling as a reliable means of achieving this effectiveness can be defended in terms of deciding what step to take next within an analysis of the constraining effect of that step upon future attainables but not in terms of constructing an ultimate called 'optimality'.

## The Art of Planning under Uncertainty

With long-range planning problems ill-defined due to uncertainty, the appropriate strategy would seem to be one of proceeding on the basis of statements of forecasts and goals concerning the direction in which investment decisions ought to move the system in order to reduce tensions which may lead to future problems, instead of on the basis of objectives precisely describing an evolution of the world in which those tensions have been removed. Therefore, an element of a system is meaningful if it can enter into the making of a decision and if its meaning is analysable in terms of the difference it makes to the decision taken. Thus, models are built incorporating characterisations of such elements to aid in the evaluation of resource allocation alternatives. As with the earlier observation on the impossibility of perfect knowledge about the future, this argument should not be taken as an excuse for myopic planning. Rather, it contends that only the goals of the present society should influence the decisions taken by the present society and that these should not be forced upon future societies. The overriding goal should be to allow the future to achieve its version of optimality; and this goal can be meaningfully and satisfactorily incorporated into present planning and investment decisions.

It should be emphasised at this point that the separation of system problems into modelling, analysis, estimation, and control is done only for ease of discussion. The circumstances of life present a mixture of all of these. In application, the various areas become so intertwined that it is impractical, if not impossible to effectively separate them. Furthermore, the planning process is iterative, usually beginning with modelling, proceeding with estimating and analysis, then returning to modelling when the analysis uncovers weaknesses in the model which may be detrimental to control determination. But, as the planning horizon increases, the unknowns become dominant and the decision-maker must exercise early judicious control to insure that those concerned can benefically adapt to the environment which actually evolves.

In noting that the use of mathematics in the social sciences 'has not achieved very much', von Neumann and Morgenstern (1944, pp. 4-5) observe that 'underlying vagueness and ignorance has not been dispelled by the inadequate and inappropriate use of a powerful instrument that is very difficult to handle'. The following chapters will concentrate on specifying the 'underlying vagueness and ignorance' and handling it with a conceptually simple, yet hopefully 'powerful instrument'. The uncertainty element is exactly the reason why forecasts and modelling are necessary and why simplicity in modelling and concentration on (but not preoccupation with) short-term decisions minimise the cost of erroneous forecasts.

The essence of the recommended approach is to provide the decision-maker with inputs to the overall investment decision. It is only assumed that the decision has been made to decide to invest. Modelling need not, therefore, be constrained by budgetary limitations, which are crucial to the successful interpretation of traditional programming approaches. These assume that the decision to invest to a given extent in a given subsystem of the economy relative to other investment opportunities in the other subsystems has been made irrevocably. The use of the outputs of modelling as proposed here enables an evaluation of the relative importance of the particular subsystem in achieving not only its efficient operation but also its contribution to achieving goals defined in other subsystems with which it interacts. For example, a regional transportation network not only strives to satisfy the general demand for movement but also provides accessibility for the development of subregions to desired levels. The central problem is that the optimising of the long-run performance of the particular subsystem is not done in a vacuum or within a perfectly controllable environment; system optimality must be defined with reference both to evolving objectives and constraints and to interdependences with the other sectors of the total economy.

The approach espoused here is not chiefly concerned with the techniques used for the actual construction of future scenarios. Modelling comes into the planning process to answer the question: What is the best initial commitment to make which will be satisfactory in the short-run and which will maximise, in some sense of the notion, the possibility that long-run optimality is attainable? Supported wherever possible by objective data and behavioural functions, modelling proceeds to evaluate the extent to which an initial investment constrains the future's options and satisfies the present's desires, operating in both instances to judiciously employ resources in the face of uncertainty.

Faced with this charge, we turn our attention next to elaborating a decision framework within which to analyse the complications of uncertainty. This will provide a firm basis of understanding of the extent to which the assumption of certain knowledge jeopardises the conventional notion of optimality in the procedures for making infrastructure investment decisions to be reviewed in Chapter 5 and road planning decisions reviewed in Chapter 6. From this we can proceed to examine the formal approaches to the handling of uncertainty, culminating with the concept of robustness as a decision criterion. This will be followed by an operational application of this approach in the context of a developing region's need for transport in Chapter 8.

# 4 ON UNCERTAINTY

It has been demonstrated that the fundamental problem of planning is uncertainty. Coping with uncertainty, then, is the essence of responsible decision-making. Since the result of decision-making is the act of control, the notion which will be explored next is that the intent of control determination is not only to guide the system toward preferred behaviour but also to reduce the uncertainty that the system will in fact exhibit such behaviour. It is this idea which underpins any discussion of decision-making, whether long-range or short-range. Although usually implicitly assumed, this is the *sine qua non* of meaningful planning. Without due consideration to reducing uncertainty, planning is conducted under total ignorance, characterised by a series of apparently random moves, and exhibiting an instability which renders vacuous such concepts as society, policy, and decision. Therefore, the structure of uncertainty needs careful exploration.

Measurements always contain an element of error. The most exact description that may be offered for any physical item is still only an approximation; and this is all the more true in socio-economic quantifications. If it should happen that there is perfect correspondence between prediction and observation, it must be considered accidental, attributed to chance, and 'should give rise to suspicion rather than to satisfaction' (Jevons, 1877, p. 457). Statistics is the theory of the structural characteristics of error; one of the functions of statistics is to provide ways of taking error into account when dealing with measurements. The probabilistic foundations of statistics may be employed to guide in the cautious, qualified formulation of quantifiable control determinations based upon data which is in some sense incomplete.

Even if a particular socio-economic measurement were free of bias and wholly based in rationality, replications of the measurement would almost certainly fail to always yield identical values. Both the prevailing concepts and contexts within which the measurements are conducted are sufficiently open to permit different observers possessing somewhat different conceptions to view what may be termed the 'same' situation. To permit the results of such multiple, subjective measurements into an objective assessment of future system performance, it is necessary to provide some degree of constancy in the construction, interpretation, and use of subjective assessments. In this way statistics performs a


DOI: 10.4324/9781003182993-4

second function; indeed, statistics may be defined as

> 'the art of dealing with vagueness and with interpersonal difference in decision situations . . . Statistics is largely devoted to exploiting similarities in the judgements of certain classes of people and in seeking devices, notably relevant observation, that tend to minimize their differences' (Savage, 1954, p. 154).

A number of observers, each making his own estimate or a single observer making estimates on successive occasions, provide a multiplicity of results which need to be reduced to some underlying unity or at least to a less divergent set of observations in order to promote consistency in the decision-making process.

In the specific case of future behaviour, such multiplicity derives from the very nature of the subject and only indirectly from the observations made upon it. Inquiry into the future performance of a socio-economic system is directed to a large set of entities, activities, inter-relationships and environments. Employing various assumptions on the nature of this set, a large set of models may be constructed, all of which are sufficiently alike to be grouped into the single model class 'possible futures', but none the less different enough from one another to be uniquely distinguishable. Whatever conclusions are finally reached as a result of this inquiry will hold, at best, for only a small subset of the model class, not for all members, and even for those for which the conclusions hold they will, in general, only hold to varying degrees of approximation.

Accepting that approximation is the best that can be obtained from observations and especially judgemental observations upon the nature of the future, resort is made to the dictum Occam's razor: 'Things ought not to be complicated except out of necessity.' This principle suggests that models should be sought to explain a system's behaviour which possess as few restrictions, in the form of assumptions, as possible upon the functional relationships. Weaker hypotheses, of course, expand the variety of behaviours admissible for analysis by a given model, thereby enlarging the class of futures whose behaviour can be described. Thus, increased applicability is added to the theory embodied in the model. However, greater generality does not come free of charge: the weaker the assumptions, the more situations that are capable of inclusion but the less that can be said about any one. Based upon the amount of information available and the model's intended application, the extent of accepted assumptions must reflect the trade-off between general and specific.

Use of Occam's razor may also be justified on empirical grounds, for existing data on socio-economic systems is not sufficient to guarantee an adequately detailed description of the mechanics of future behaviour. As future data becomes available, narrowness of hypotheses and the resulting strictness of model implications can only make rejection of the present decision's optimality all the more likely. Thus, it is worth considering increasingly general views of the future which, by virtue of their broader scope, have a better chance of selecting a control which will benefit the future which actually evolves. Here, again, a trade-off must be considered since present decision-makers should attempt to maximise the future effectiveness of the applied control, which if too general may not achieve its maximal effectiveness.

Here, then, is the question to which this discussion is addressed: Can the multiplicity of observations be reduced to some unit with respect to which investment sets may be evaluated as to their desirability? What is the smallest set of descriptors which will best characterise the heterogeneous class of possible futures, and how good a description does it provide for the purposes of transport network investment?

Suppose a model is presented which is intended to cover a large class of systems and to describe or explain a large class of evolutions. By the arguments of the previous chapter, the evidence supporting the model can never be absolutely conclusive and the level of confidence in it can never attain complete certainty. The evidence supporting models of an evolving system will be taken from only some of the instances within the range of any conclusion, that is, historical data; a degree of uncertainty will be attached to the remaining cases, that is, the unobservable future. Even where the historical data are exhaustive (where *every* measurement has been made), the element of error in each datum cannot be left out in analysing the weight of evidence in support of any conclusion derived from a model using the data. What is worse, it is likely to happen that while some of the instances support a given conclusion, others run counter to it; and the positive evidence must be weighted against the negative. These considerations suggest that statistical concepts may also be employed to assign some degree of confirmation or of credibility to the conclusions resulting from the analysis, in the light of the inconclusive evidence on which such conclusions are always based in the interpretation of socio-economic system reaction to applied controls. Statistics, then, may be said, in the present context, to consist of ways of dealing with a multiplicity of data so as to determine what control determinations they support, and how much support they provide for each toward the attainment of obectives. From this standpoint, statistics may be characterised as dealing with or quantifying the reasoning process employed to make decisions in the face of uncertainty.

Knight (1921) established the simple and oft-used trichotomy of certainty, risk, and uncertainty to characterise decision-making modalities. Here *uncertainty* will be taken to subsume all three, and, to more accurately label the last of Knight's trichotomy, *ignorance* will be employed. In Chapter 3, the nature of uncertainty within a system was explored; uncertainty will now be examined in the context of probabilistic assessments of future realisations stemming from the causal texture of evolving environments.

*Certainty* is the degenerate case of uncertainty, wherein it may be asserted that the future is characterised by a sequence of events each of which carries probability unity of occurrence. Methodologically, the dichotomy of 'a certain event' (which will be referenced by the term 'determinate') and 'an event with probability unity' may be disturbing, for within long-range planning considerations there can never be an event such that every other mutally exclusive event may be labelled 'impossible'. Philosophically, to consider some event 'determinate' presents some obvious complications for the whole question of choice and control, and to act as if the event were determinate or certain is to ignore the consequences of being wrong. The best that may be hoped is to select controls in such a way as to render all undesirable events infinitely improbable. It may then be said that the remaining set of reasonably probable events are those to which the planning process should address itself, for one of their number may be taken as certain.

*Risk* characterises a perception of the future when there is knowledge of the law which governs the realisations of events but this law contains a random element of known behaviour. The probability is unity that the probability density function over all events is known. That is, the probability that a given event has a given probability of occurrence is unity, or equivalently, the probability of each event is given but there is ignorance about what the actual outcome will be. (This modality of decision-making is that generally inferred by the term 'uncertainty' in the planning literature.)

*Ignorance* is characterised by not knowing the probability law which may be operating. Although there need not be ignorance of all the circumstances involved, not enough is known about all the mechanisms and relationships to assign determinate probabilities to the events in the set of perceived possible outcomes. This modality more accurately reflects, in varying degrees, the true nature of the long-range transport planning problem.

It might be argued that the difference between the three types of uncertainty is not a fundamental one: a planner is never wholly ignorant of relevant probabilities; but risks are not so determinate as might generally appear. Although a set of events may be selected such that it may reasonably be asserted that the future will be found to be in that set, it will never be known, even after the fact, what the probabilities are; only estimates of these probabilities can be made, and these estimates will be more or less in error. Indeed, estimates are all that there can ever be, for if the nature of the error were known, if not in magnitude at least in direction, then better estimates could be made, but they would still be estimates.

The distinction between certainty, risk, and ignorance, though philosophical, is of great methodological importance. Not infrequently, especially in long-range planning literature, situations replete with ignorance are treated as though they were accurately representable by determinate risk formulations. Philosophically, it may be argued that probabilities of the future do not exist, but methodologically the construction of logically consistent probabilities to guide in the selection of controls is meaningful. What is objectionable about such constructions is not the making of assumptions that go beyond what is known – all assumptions are of this character. The objection is that there is ignorance as to how much is being assumed – what is the nature of error and how might it be diminished, if not in extent, then at least in effect? To contend that the future may be approached probabilistically is to acknowledge that the future derives from the interactions of entities whose behaviour either is probabilistic or is only known probabilistically.

Given the probabilistic operation of a system, controls are selected and applied; but, not only do the controls possess probabilistic properties, the system's reaction function is probabilistic. Since there is ignorance as to the exact form of any of these proabilities, assumptions must be made in order to assign measures to futures that are amenable to manipulation within the calculus of probability. This ignorance is usually made the occasion for a pretence of knowledge rather than the justification for further inquiry. If the probabilities are not known, they are not known. Ignorance is not a premise from which it may be concluded that the elements or a set of mutually exclusive events are equiprobable or anything else.

The danger of certainty and risk formulations lies in assuming that sufficient evidence exists in a particular case to support a probability assignment without being aware even of what all the necessary premises are that would justify the realisation of the associated future. In the strongest sense, to regard a future as certain is to discount the risk of error in acting on its 'truth'. In a weaker sense, to regard some probability measure as certain and employ an aggregation method over such measures on various futures which does not preserve explicit consideration of uncertainty presents equivalent difficulties. Ignorance is bliss until some future society, which very well may contain many members of the present, must pay the price for the mistakes.

**The Search for Optimality**

The determination of an optimal sequence of controls to be applied to an evolving system to insure the attaining of an optimal sequence of states would be aided immensely by a means to predict accurately the state into which the system will evolve. This, in turn, would be immensely simplified by dealing with a determinate system! Fixing the present circumstances of a determinate system determines the state it moves into next. Furthermore, since such a system cannot occupy two different states at the same instance, the transformation is unique. Of course, fixing the present circumstances of the system requires that the variables and relationships be few enough for human comprehension and that control may be exercised over all components or that those over which control cannot be exercised exhibit behaviour which may be reliably predicted. In other words, the requirement is for a closed system or, if closure is incomplete, for predictability of exogenous forces acting on it.

There are two observations modifying this requirement. The first lessens the necessity of control over every variable by noting that it is the behaviour of the aggregate system within the authority of the planner which is evaluated. Then, within a given system, the variances among the predictions of individual component's behaviour is lessened by consideration for the behaviour of the aggregate system. Therefore, since the effectiveness of a control is measured with respect to system behaviour, which being an aggregate possesses some inertia which renders it more predictable than its individual component's behaviour, the desirability of a control is more easily evaluated.

The second observation emphasises the need for identification of the individual component's states and controllability by noting that the behaviour of a system is evaluated in many different ways, by many different observers possessing many different concepts of what constitutes a 'good' system. Here, again, unity is sought to guide control selection. With respect to the system's behaviour, there will be those who will say that it is 'bad' and those who will say that it is 'good'. This divergence results from differing amounts of information among the observers, their viewpoint with respect to the system, and differing value judgements with respect to the various components comprising the system. It is easily postulated that no socio-economic system within our present or foreseeable comprehension might be construed as 'perfect' by even one observer, much less by the entire constituency, and, therefore, that any system may be improved. To optimise the effectiveness of the allocation of resources it is necessary to partition the system into sufficiently small components that unanimity among the observers is attainable as to the desirability of the allocation among partitions. Here, effectiveness is construed as reducing those tensions which the observers perceive as plaguing the performance of the system. The essence, then, is to identify the ingredients of the system that are necessary contributions to society's goals, to ascertain to what extent each contributes to society's well-being, and to structure an allocation of resources to attain highest efficiency in attaining those goals. But this strategy does not avow the inherent uncertainty of the future, it only translates that uncertainty into a collection of majority opinions as to what should be done now to achieve future optimality.

It may be contended, in the manner of Laplace, that the system is determinate by nature and that it is only an incomplete understanding of the mechanisms which allow the intrusion of uncertainty. None the less, it is valuable to consider the planning process as consisting of a sequence of control selections within a dynamic socio-economic environment which does not always fully disclose the alternative actions available or the consequences resulting from the chosen courses of action. That is, the planning agency does not always have full knowledge of its present options or the options that will be available in the future as a result of its present decisions. Then, the agency has limited capacity to gather and process information and to predict consequences of possible controls: to face the full complexity of uncertainty would overwhelm the planning process, bringing it to a full stop. Therefore, societies 'adapt well enough to 'satisfice'; they do not, in general 'optimise'' (Simon, 1956).

Unfortunately, the traditional systems of epistemology do not contain the necessary concepts for simultaneously thinking about rationality and indeterminateness; these appear to be incompatible concepts. There is no ready way of thinking about something as half-closed, half-rational. The simplest and most often abused mode of structured planning is to employ the completely closed-system approach, ignoring uncertainty and claiming pure rationality. This is the foundation employed in the application of deterministic optimisation methods to transport network planning. Another, and perhaps more potently destructive, alternative, is to ignore rational action and proceed spontaneously. In this discourse, society is conceived of as an open system, hence faced with ignorance, but at the same time subject to some criteria of rationality, hence needing determinateness or at least some degree of certainty. Consequently, then, the traditional question of planning theory 'What is the control?', which begs for an authoritarian answer such as 'the optimal one', is recast as 'How can we organise the planning process so that "bad" decisions (which we should try to not make, but whose unforeseen consequences may be wholly unsatisfactory) cannot do too much damage?' or as 'How can we hope to detect and eliminate the risk of error in acting upon our perception of the future?'

**Control of Evolving Systems**

It follows from above that societal goals are established through a coalition-type behaviour. Therefore, decision-making always possesses two basic major dimensions: (i) beliefs about the causal texture of the system's environment, the nature of cause-effect relations; and (ii) preference orderings among possible outcomes. These are the basic variables of decision-making. Although they are not always explicitly introduced, they are none the less present. The first dimension is the arena of modelling and its successful treatment supports decision-making in the attainment of the second dimension's structure. Each of these dimensions may be trichotomised consistent with the present view of uncertainty, thus yielding nine types of decision-making environments. It would seem clear that each type would require a different decision-making strategy which would reflect the varying importance of estimation and control as discussed in Chapter 3.

When there is certainty regarding both cause-effect relationships and preference orderings, the strategy is purely computational. 'Optimality' is approachable within the situation but may be unreachable and certainly cannot be guaranteed since all options may not be known or there may not be enough computational power available to analyse all the options. Within such a certain environment, the limitations upon optimality become physical and not philosophical. For example, consider the determination of the 'optimal' set of tours for newspaper delivery in a large town. By the time the optimal could be guaranteed, all the presently available newspaper boys would be dead, although the set from which the optimal must come is knowable exhaustively. This point is consistent with the notion that 'a decision delayed becomes no decision' and suggest that in some cases too much may be known about the extent of options but not enough about the properties of the options themselves. In essence, although the system's causal structure is certain and there is certainty regarding the preferred outcomes, the most efficient decision is not always obvious.

Where outcome preferences are certain but the texture of the system is represented by risk or ignorance, judgement characterises the approach to decision-making. This would seem to describe the general societal context in that there is general agreement that 'things' should be made 'better', but there is basic disagreement as to what this means exactly with respect to exactly what should be done to achieve such betterment. Where the situation is reversed and there is risk or ignorance attached to the preferences and certainty with regard to the responsive behaviour of the system, decision-making consists of compromise among the preference factions as to what should comprise the system's goals. Given a viable compromise, decision-making becomes computational. In the case of risk or ignorance about both dimensions, decision-making takes on the form of inspiration if it is to be meaningful, otherwise the system moves at random or spontaneously with nothing more than the instantaneous societal goal of survival.

The recognition of uncertainty of preferences among outcomes identifies a key problem in the process of planning. On the one hand, it may represent an incompleteness of knowledge on the part of the decision-maker regarding the preference orderings within his realm of control. Even though all the variables are under the control of the decision-maker, lack of knowledge of the present prevailing preferences calls for a judgemental strategy. The more likely case, however, is that not all the variables will be under the agency's control, in particular, those related to the behaviour of the individuals within the system. Then, the effectiveness of any applied control depends in very large measure on how well it is received by those individuals, how the system reacts.

On the other hand, and more importantly, uncertainty about preferences among outcomes must be viewed with respect to the value systems which prevail when these outcomes are realised. When compromises are sought, they operate on two levels: (i) among divergent factions within the present and (ii) between the present, who possess the power of control, and the future. Even if there is certainty about present preferences (i.e. everybody agrees to the present goals), there must still be compromise with the future, for, although the present is not wholly ignorant of the character of the future, it is sufficiently so as to necessitate compromise. Even complete knowledge of the future does not negate the need for compromise if the future stream of resources to be allocated is finite.

The compromise between present and future preferences may be achieved operationally in investment planning by maximising the flexibility of response of the infrastructural system in question, e.g. the transport network. Flexibility represents a concession to uncertainty and an abdication of optimality in favour of performance not being too suboptimal if the conditional premises upon which optimality is construed were not to materialise. Flexibility embodies compromise as an increased cost of investment by the present to gain the level of performance it desires while enabling the future to expend less than it would otherwise to adapt the subsystem to its desired level of operation.

The following example of real-time control can be taken as a parable for long-range decision-making. A motorway equipped with surveillance instruments can close on-ramps to traffic in order to produce flow stability on the motorway. The addition of surveillance to the motorway enables adaptability of entry conditions to congestion conditions by producing a source of knowledge available to the decision-maker on the need for controlled entry to maximise the performance of the motorway in moving vehicles. The flexibility of the system is called upon to reduce the flexibility of entry under the knowledge that the latter flexibility will be detrimental to the otherwise optimal performance of motorway. Flexibility may be reduced in such situations since knowledge of the state and the probably dynamics of the system exists – 'probable' being employed to recognise the possibility of, say, an accident disrupting the flow.

Equivalently, the decision-maker in long-range investment situations may reduce the extent to which he incorporates flexibility into the system as he possesses more knowledge of the future environment within which the system will operate. If the decision-maker is certain of the environment then he may seek the 'optimal' investment in the traditional sense. It is the more general case, of course, that the decision-maker is always looking for 'the' optimal investment and, as uncertainty increases, flexibility becomes a very important component of optimality in its fullest sense. The need for flexibility, then, is a direct function of the extent of uncertainty prevailing about the future over the life of the system, which may be greater than the life of the investment in question.

Optimality may be cast in terms of minimal long-run transport costs, which consists of the present fixed cost and future maintenance and operating cost. Therefore, optimality must incorporate consideration of the future adapting the system to its needs. Then, this notion of optimality suggests that the system which satisfies a certainty perception of the future is optimal but, as this certainty tends to evaporate, greater and greater flexibility must be provided. This is especially important with respect to societal values, for if an inflexible system is put in place the future must either operate within the present values or scrap the entire system in order to properly pursue optimality. The former case implies a burden upon the future and the latter a burden upon the present who could have invested in a shorter-lived system at lower cost. Any position which is not coincident with prior value systems implies an increased cost of utilising prior investments to remain consistent with the new position.

In the context of evolving utility functions, the search for optimality, however it is defined, and the concomitant resort to rationality, are both couched in terms of consistent behaviour in the same manner as Koopman defended the axiomatisation of subjective probability:

> *A given individual at a given moment* may be regarded as assenting to a certain set of propositions; ignoring what he may hold at any other moment or what others may believe, that set of propositions which he holds at that given moment must have certain relations with one another which may be called *relations of consistency* (1940, p. 163).

This set of propositions comprise the support of the chosen investment and it is assumed – explicitly or implicitly – that the truth of these propositions will remain stable over the relevant horizon required for the investment's justification. Therefore

> the statement that 'given a set of feasible alternatives, the "reasonable" [rational] man should choose the best' and the parallel statement that 'an actual man does choose the best (and if the best is not feasible, the second best, etc.)' are both empty if the ranking order is not supposed to be stable over a period long enough to make the statements of practical relevance? (Marschak, 1968, p. 43).

If the ranking order may be construed as stable, then the practice of decision-making as the selection of the best controls becomes possible. It is the enrichment of this premise within the planning process to which the discussion now turns.

**Facing Uncertainty in Decision-Making**

It has been the intent of this chapter to seek a logical foundation for a rational approach to selecting investment commitments in transport networks which will yield short-term benefits while not constraining the ability of future decision-makers to achieve equally satisfactory results. In order to do this it is necessary to demonstrate that the espoused approach provides for the systematic treatment of the uncertainty inherent in both the infrastructure involved and its controlling environment. This boils down to uncertainty over the future evolution of spatial demand patterns and over the ability of the resource allocation process to provide proper levels of capacity at the right place and time, efficiently.

The roadways of a transport system require a long planning horizon due to the length of their expected service life and the long period needed to recover in benefits the large amount in committed capital resources. Within this class of locationally-fixed capital ventures, the planning and design of roads for developing areas provides a good framework for analysing the meaningfulness and efficiency of the proposed methodology; and this will be undertaken in Chapter 8. In particular, a developmental context demonstrates more easily than a developed context the constraining nature of initial major investments upon the extent of acceptable investment alternatives available to future decision-makers. The transportation infrastructure of developed areas possesses an inertia from its history that clouds constraining effects of new investment; of course, this inertia embodies the constraining influence of the investment history upon present alternatives. Although a developing area possesses an operating spatial economic pattern and attendant potentialities to which the planning of any new network investment must be sensitive, the new investments are of a magnitude so much greater than the existing system that the latter's constraining effects are relatively minimal.

However, the considerations explored here are generally applicable within the decision-making structure of the class of physical system which requires large investment in fixed plant and which require long horizons with uncertain *a priori* estimates of demand and technologically feasible capacity levels for their economic and social justification.

Examples of such possible applications are in the planning for urban rail-transit systems, pipelines, electric power networks, communication systems, aqueducts, and industrial and public facility location decisions, as well as expansion of networks within a mature economy.

Underlying the decision-making structure for such systems is the dichotomy between the visionary requirements of long-range optimisation and the political exigencies of getting something done now: between theory and practice. This conflict derives from at least three temptations plaguing the planning process (Heymann, 1965), each of which may be viewed as a response to a type of uncertainty in economic evaluations and measurements (Morgenstern, 1963): (i) The temptation to pursue the answer in the name of efficiency as if there were no irreducible elements of uncertainty in the subjective formulation of goals. The risk here is that the planner will bias his evaluation of society's desires by his own views of the nature of social welfare. (ii) The tendency to justify investments mainly because they are supported by quantification. Such supporting numbers can be questioned as being derived from uncertain inputs, both as to whether all necessary information has been gathered and whether its quality has been evaluated correctly. That long-range planning is the analysis of decisions under uncertainty of quantitative guidance is almost definitional. And (iii) the temptation to ignore immeasurables or intangibles in the analysis and decision-making process, to exclude important complicating factors: to oversimplify the complexities of the real world.

Since the objectives of society are vague and incommensurable, it is difficult to guarantee that the proper conclusions have been drawn from observable phenomena so as to accurately integrate information on the phenomena into a satisfactory model of the future course of development of any given area, even if the objectives are stable. Perhaps the most serious shortcoming of modelling which results in planning errors is the incorporation of gross (and oftentimes hidden) assumptions to circumvent these difficulties. The existence of these assumptions often times makes the meaningfulness of conclusions from models hard to interpret back into the real-world societal context. Assumptions are imposed to overcome problems of observation but also for computational tractability. Regrettably, the effects of these assumptions are very seldom investigated, even though they may render the conclusions of the model invalid in application. Of course, the particular set of assumptions may derive from the state of the art of modelling. But, one suspects that a model

> with no increasing returns to scale or external economies, no monopoly elements or financial constraints, no differential social welfare effects, and so forth – and with perfect knowledge – . . . [is constructed so that it becomes] a simple matter to determine investment priorities (Sadove and Fromm, 1965, p. 227).

The pretension with which 'purists' sometimes present such models to actual decision-makers conflicts with the following observation of an accomplished model-builder:

> If I were a planner, how much would I believe in the predictions of these models? If one of the large LP models gave rise to a prediction that conflicted with my intuition, I am afraid that in the present state of their development, if I found no way to reconcile the two it would be the LP result that I would throw away (Kleitman, 1976, p. 241).

A decision-making structure which does not strive to overcome the temptation of excess assumptions or to reconcile its divergence with professional judgement proceeds as a basically subjective process, in spite of any superficial appearance of rational elegance.

The essence of this problem revolves around the failure of social scientists to develop a comparable tradition to that of the 'exact' sciences in exploring errors and evaluating their influences upon the decision-making process. The acceptance of the Heisenberg uncertainty principle in modern physics represents recognition of 'the highly important fact that this impossibility of certain measurements is more than a mere technical limitation that someday might be overcome by clever instrumentation' (Weisskopf, 1972, p. 32). That is, this indeterminacy is inherent in the nature of a moving body, which may be taken loosely as equivalent to a collective dynamic society. The principle results from the fact that the very act of measurement introduces uncertainty since the object responds to the performance of measurement by changing its state (Beiser, 1967). In the same way, the act of assessing the past performance of a forecaster affects his future performance (Schlaifer, 1969). For example, a forecaster may have presented consistently lower estimates of future demand than were actually realised. The act of making this observation and making the forecaster aware of it may bias his future estimates of probabilities toward a higher mean value than would have represented his forecast otherwise. This supports the argument that probability assessments of the future are things that someone thinks and are not things actually possessed by that future.

The intent of measurement is quantification in modelling. This observation leads to the First Fundamental Problem of Measurement (Suppes and Zinnes, 1964, p. 41):

> The Representation Problem: Characterize the formal properties of the empirical operations and relations used in the procedure and show that they are isomorphic to appropriately chosen numerical operations and relations.

The goal of modelling, then, is to construct a tractable system using measurable phenomena which is isomorphic with relevant systems in the real world. In modelling endeavors, it is generally not possible to demonstrate that a discrepancy between theory and observation is attributable to data error rather than to an inadequacy in the theory around which the model is constructed and the measurements conducted. Similarly, it is not possible to attribute failure of an investment sequence to achieve optimality to either the structure of the model or the accuracy of the inputs, for, indeed, future socio-economic environments can neither be *a priori* observed nor measured.

Therefore, modelling must only be employed as the guide of action, since 'the truth which it arrives at is not that which we can ideally contemplate without error, but that which we can act upon without fear' (Clifford, 1872).

In a model-based framework of planning, both decision-makers and model-builders are operating within a process which may be viewed in a system-theoretic context, with uncertainty in its same complicating capacities as before. Briefly, this system-theoretic view consists of an input-out relationship between the two teams: decision-makers attempt to estimate societal values as input to the model-builders, who try to model society and analyse consequences for the decision-makers, who then attempt to determine proper controls. Each can make incorrect evaluations within its own realm, but the effectiveness of the entire process lies solely in the future performance of the implemented action in satisfying demands made upon the system. With respect to the relationship between the actual commitment and the society which actually evolves, the errors possible in control implementation may be likened to the Type I and Type II errors of statistical hypothesis testing:

I. not adopting a plan which would have proven satisfactory; and

II. adopting a plan which proves unsatisfactory.

To elaborate, consider a historical context. If a system were now being built *de novo*, it is very unlikely that the structure of the system would be that which has actually been constructed. Plans are adopted for the purposes of alleviating tensions imposed on the present system by misjudgements of past decision-makers and providing for the anticipated needs of future users. Present tensions have been created due to a deviance from what was expected by earlier decision-makers to obtain at the present time, due to different expectations by present users from those values which satisfied the desires of those users to whom the earlier decision-makers were responsible, and due to differing views of the future; this evaluation is of course *a posteriori*. Therefore, plans were implemented based on criteria 'that have lost their significance, but which have imposed repeated problems on later generations as transport, supply, and municipal services have changed' (Sauer, 1941, p. 374).

Again, the intention is not to suggest that nothing should be done until the future is known with certainty. Indeed, it is not so much a problem of deciding not to adopt a certain plan because it may fail to prove satisfactory to future users, but that indecision may prove to be the real problem in a growth environment. Delays in adopting a plan because it is not known which alternative will prove optimal introduces delays into the consideration of alternatives and has significant impacts on the welfare of the users and the ability and relevance of those alternatives to improve the standard of living. This may be interpreted as 'a decision delayed becomes no decision', since by the time a plan has been studied and approved by everybody and is deemed optimal, the criteria and circumstances upon which it is founded may no longer be valid.

Therefore, the dictates of reality require that planning with imperfect knowledge in an imperfect organisational and institutional context should concentrate on what can be done now in light of what has been done in the past and with appropriate regard for the future, and not what should be done in light of what someone thinks will happen in the future. This represents a change of emphasis from planning the total sequence of decisions subject to criteria which temporally aggregate performance measures over the planning horizon or which are conditioned on some overriding notion of end-state optimality or long-range goals. Placing emphasis on selecting the most favourable initial commitment aims at minimising the degree of risk that the evolution of the long-range investment programme may prove unsatisfactory. Such a planning focus maximises the potential effectiveness of future adaptive controls in achieving near-optimality within the actual evolution of the preference-controlling environments.

This position may be demonstrated by analogy to a physical horizon. Consider a person setting out on a journey on the ocean with the objective of reaching some port a long way off, beyond the horizon. The navigational decisions that he will be required to make beyond the horizon will be unknowable in detail, but he will start the journey with every confidence that the proper decisions can be made as problems arise. His initial problems are first, to start the journey at all, and second, the detailed decisions associated with the pre-horizon conditions that are perceived and can be planned for. As the journey progresses, the ability to evaluate the influence of the horizon varies as a function of the journey's course and the state of the environment, with increased predictability when clear and calm, and with increased necessity of short-range control – and perhaps destination modification including a return to the initial port, if the initial commitment is not irrevocable – during natural adversities or as a result of the functional loss of his navigational equipment.

This planning horizon concept may be exploited in much the same way for large-scale investment decision-making. Post horizon goals must be weighed against possible costs incurred due to the uncertainties of welfare measures upon future users. The detailed investment commitments must be made within the pre-horizon period in concert with post-horizon goals. Therefore, the goals must be defined with a high degree of confidence that they can be achieved and that such achievement will be desirable by future users, i.e., at the time the goals may become reality. As time progresses, more and more commitments will need to be made and each will depend on the evolution of society and the continued enabling of the achievement of those goals. And, as society evolves, the desirability of those goals must remain, for each goal-oriented investment is not without its real costs. Of course, the time will come when a society will endure those goals as realities and the attainability of that society's goals will be dependent upon the constraints imposed in achieving the intially-formulated goals. Therefore, careful and detailed planning is needed for pre-horizon investments even though the go-ahead on the decision to invest at all should not be and really cannot be based solely on pre-horizon considerations.

Of course, in the same way that the navigator above may attempt to return to port if conditions prove too adverse and he is not too far committed, the decision-maker may attempt to return the system to a prior state should the application of a control lead to performance less satisfactory than that which prevailed before the application. A simple example of this would be a highway approach ramp which generated more congestion on the highway than it relieved on the arterials: the solution being to seal off the ramp, at least until a further study can be conducted. For both the navigator and the decision-maker, the experience and knowledge would have been costly, especially as the time and funds could have been invested in potentially more productive ventures.

## Conclusion

Hopefully, this chapter has demonstrated the näivety of deterministic choice rules based on assumptions of certain knowledge of preferences and processes. A general case has been made for changing the focus of the planning perspective so as to emphasise current needs, while catering for future needs by building flexibility into structures. This avoids the obsession with an imaginary singularity with characterises much applied optimisation and planning. Before developing and applying an appropriate rule of choice to embrace the reality of our lack of knowledge of the future, we need to show plainly what is involved in assumptions of perfect knowledge. This is done in the next chapters by reviewing the rubrics of decision under certainty first in general and then in transport network terms. Doing this with the above discussion in mind should put us in the position where we can begin to dismantle the specious simplicity of these procedures.

# 5 DETERMINISTIC OPTIMISATION

Prior to getting down to the specifics of road planning under assumptions of certainty, it is appropriate to set this problem in its general methodological context. To this end we will review some proposals for the application of deterministic techniques to investment decision-making for locationally-fixed, widely-developed structures. These structures include facilities for the movement of telephone calls, water, energy, passengers, and freight.

The planning problem within such an infrastructure lies in the long construction period necessary for projects whose effectiveness is essentially all or nothing in nature. Either a dam exists or it does not exist. If it exists, it may be in the 'filling' stage and, although able to control flooding partially, it is unable to increase the amount of water available for irrigation and pollution dilution in low flow periods significantly. Of course, even with a completed dam, design considerations and Nature are such that a possibility of uncontrollable flooding or unabatable drought always exists.

To be sure, the initial explorations of the planning of any system would ideally admit every conceivable possibility of achieving the formulated goals and objectives. However, that set of commitments which actually constitute the final realisation of a planning endeavour is not selected from an infinite population of alternatives. At the start of a planning venture certain sites which are amenable to the construction and operation of a dam or a hydro-electric plant, certain routes which are feasible for the construction of a highway link, an aqueduct, or a natural gas pipeline can be identified; and it is these alternatives from which the actual commitments will come.

The designation of these admissible points and lines derives from the existing technology and the existing spatial distribution of resources and society. The structure for decision-making established in Chapter 4 seeks to determine what investments are to be committed to now. Therefore, existing technology necessarily determines what is technologically feasible for present consideration. Further, such feasible locations admit only a small number of technologically realisable constructions. It is begging the question to contend that such items as carriageway width are continuous and, therefore, that an uncountable infinity of widths are admissible. This bounding of admissible


DOI: 10.4324/9781003182993-5

alternatives is not a simplication to achieve mathematical tractability but a construction reality.

Geography exercises a strong, indeed primary influence upon admissibility. For systems related to natural resources, the availability of such resources is crucial. For example, a sculptured mound of dirt is a piece of engineering art if it does not retain water. The geography of demand for some item is related to the item's past availability and potential for future availability. For example, a natural gas transmission pipeline necessarily assumes a source of natural gas and relies on there being sufficient demand for the product over sufficient time to recoup expenses. The northeastern United States would not have demand for natural gas had its industry not had gas available during its formative years; therefore, the actions of past decision-makers contributed to the creation of the present demand for natural gas and imposed upon present and future societies the increased burden of higher-priced natural gas or the immense expenditures to retool affected industry. Further, what may be characterised as the decisions of an even more temporally remote decision-maker, Nature, dictates the location of potential supplies. Hence, the spatial organisation and the technological ability of society restrict the real extent of consideration for natural gas pipelines.

The restrictions stemming from the limited abilities of present society and societal characteristics stemming from the decisions of past societies do constrain present decision-makers. Aware of the restrictive effects of past decision, present decision-makers should consider potentially similar effects when making large-scale commitments which will restrict the alternatives open to future societies. Continuing with water resource examples, to contend that the present withdrawal of groundwater should be continued for some region until it is exhausted – thereby delaying the decision to invest in a large-scale water transfer project to await the possible economic feasibility of desalting plants (as suggested for northwest Mexico by Cummings (1974)) – provides restrictions for the future just as potent as if commitment were to be made for a transfer project. Although such delay allows time 'for considerable research on alternatives for increasing water supplies' (Cummings, 1974, p. 109), it also imposes upon the society of the near future an increasing anxiety with respect to demand increasing toward and beyond the recharge capabilities of the supporting aquifers. Such anxieties may be reflected in a lesser degree of regional industrial development than would have occurred had the project been undertaken. This lesser development and associated necessity of increased efficiency in water usage would be the result of the applied control 'continue with existing sources' and would constrain the development options open to the region.

Of course, commitment to a large-scale water resources project tends to have irreversible effects upon the natural environment and rationality may suggest that it is easier to build a dam later than it is economical to not use one once it is built.

> Long life (current calculations of benefits and costs [for water resources projects] are typically based on an expectation of 100 year life) means a lack of flexibility in incorporating technological innovations in the water system [if it is built]. A desalting plant or a water reclamation system with an economic life of twenty or thirty years may be disassembled and replaced by new equipment if technological improvements warrant such a change, but once the storage and transport facilities of a [water] transfer system are in place, it becomes physically difficult to remove the structures and probably impossible to regenerate the natural ecosystems in the reservoir areas. In general, longer life means greater uncertainty – economic, technological, and ecological – against which the investment bet is made (Howe and Easter, 1971, p. 5).

Aside from the observation that thirty years to a single-product unit operation sensitive to technological change cannot be defended as incurring meaningfully fewer uncertainties than one-hundred years to a modified natural system with many products not so sensitive to technological change, consider the basic economic rationale for investment in either item. A desalting plant would be installed if the future returns from its operation were considered sufficient to call forth present capital resources. This must also be the basic principle governing the decision to invest in a dam. If technological change renders an in-place operating desalting plant obsolete, it is highly unlikely that it will be desired elsewhere. If every place has access to the same type of new plant which is considered economically justified to replace the old plant, then any place requiring a desalting plant will most likely also opt for the new plant. This same notion of technological obsolescence operates with a dam. Therefore, both will deteriorate in-place should technology render either obsolete. 'Investment is, in essence, *present* sacrifice for *future* benefit. But the present may be relatively well known, whereas the future is always an enigma. Investment thus commonly also represents *certain* sacrifice for *uncertain* benefit'

(Hirshleifer, 1970, p. 215).

One certainty associated with uncertain future benefits is the deterioration, either in a technologic/economic or in a performance manner, of the ability of components of physical systems to recover benefits. Desalting plants break-down and require more maintenance cost with age; dams silt-in, providing less flood control with age. Consideration for performance deterioration is conspicuous by its absence from mathematical programming approaches to investment in locationally-fixed, large-scale systems. The importance of this omission is that, although the present value of utility in distant futures is small, the influence of deterioration upon attainable future utilities of large-scale present investments will not be insignificant. The implication, then, is that either (i) present decision-makers do not care for the well-being of future generations, (ii) the investment is seen by the present as operating in relative perpetuity within an accurately forecast environment, or (iii) the present leaves to some intermediate set of decision makers the task of taking care of any problems resulting from the present investment before tensions develop in the future system.

As the first possible implication is philosophical and the second conceptually expedient, consider the third which is operationally expedient. Although *homo economicus* attributes (virtually) zero influence upon his present behaviour to the assurances from the scientific community that life as we know it will cease on this planet in some thousands of years, the probabilities of technological advance and social value change cannot be considered so insignificant in the planning for projects whose incorporation into a socio-economic system will necessarily influence both the operation and configuration of that system far beyond the projects' service life. Admittedly, within a complex system, such as an intra-urban transport network, consideration of major improvements, minor improvements, facility abandonment, and potential natural destruction of all interrelationships is a difficult task at best. But, is it any better to subscribe to the strong assumptions and gross simplifications necessary to employ highly exacting, though elegant mathematics to assist in the analysis of planning a complex, diversified system? Due to societal limitations, such systems admit but a finite number of realisations, possess a finite service lifetime (though infinite influence time), will be constructed over a period which cannot be plausibly construed as 'instantaneous', and will operate within a socio-economic environment about whose characteristics the present planners are uncertain.

Rather than continue in a general vein, groups of ideas which exhibit the state of the art in investment planning relevant to water resources, regional growth, and networks will be presented to demonstrate the validity of these assertions.

*Water Resources*

The considerations relevant to the steering of regional growth by water resource system investment are, in general, the same that surround the design of transport systems. However, the selection from potential projects is much more restricted in space, except for the route location of associated aqueducts and pipelines. That is, whereas a vehicle transport system in a city is controllable at many locations by such additions as traffic lights on minor streets, provision of reversible lanes on motorways, and policemen, the movement of water in a river basin cannot feasibily be controlled except at a few locations by the construction of dams and, in some cases, locks. Although the potential sites for control are more limited in water resource systems, the ecologic-economic consequences of investment at a given location are more complex and more diversified. But, as in vehicle transport systems, each location admits only a few potential realisations although the combinatorial possibilities of the technologically feasible constructions for achieving the various criteria – flood control, power, water supply, recreation, irrigation – are greater.

Since there are many more objectives to be contended with, the investment sequencing problem for a large-scale water resources project is not so easily formulated as that of a transport system. This multiplicity of performance criteria is all the more complicating in light of the relatively longer life of dams, aqueducts, and, to a lesser degree, hydro-electric plants. Due to the essentially irreversible consequences of water projects, it is of utmost concern to consider the socio-economic environment of the future and its relationship to the present; and this concern is appropriately reflected more in the literature of water resources planning than of transport planning. Ignoring physical deterioration and retirement of support facilities over a twenty-year horizon is questionable in most transport investment processes but not so critical in water systems. But being in error in forecasting dominant variables is more serious since, unlike vehicles which simply slow down and run into each other as a function of congestion, water will exact great penalties in flood damage if holding capacity or the structural factors are insufficient.

It is interesting to note that different time horizons will give rise

> to different sequences, so that adding one short-range plan to another would be more costly than one long-range plan [*if the forecasts are realised*]. On the other hand, however, demand [and other] predictions become more uncertain with further progression into the future. (Becker and Yeh, 1974, p. 61; emphasis added).

To emphasise and elaborate upon this point, the determination of the 'optimal' planning horizon for the 'optimal' selection of immediate investments must balance off the very-long-range consequences of such actions against the relatively great uncertainties associated with forecasts of relevant factors in distant futures. Such a balancing is the essence of effective planning and represents the influence of the convergent nature of an aspiration and the divergent nature of a prediction. Toward this optimal determination, Butcher *et al.* (1969) present a method of selecting initial investments which are least sensitive to variations in planning horizon length over some relatively meaningful period. The planning horizon is the single most important control that the decision-maker can manipulate to maximise the effectiveness of the planning process. Technology may be considered to be relatively fixed, demand changing in a fashion consistent with acceptable indicators, and interest rates on capital for the short-run predictable from the general business environment; indeed, the planner has little direct control over the behaviour of these factors, but he does have considerable (if not total) control over the length of planning horizon which will be used to justify his initial commitments. Further, the physical and economic interdependencies of the major project and its supporting facilities reduce the combinatorial complexities associated with each potential site that the topography permits. Now, the planning process need only determine that sequence of the selected set of projects which maximises its objectives and the initial commitment coincides with the first elements of the sequence! Using dynamic programming, Butcher *et al.* allow the planning horizon to increase from some minimum that yields a non-trivial solution to a maximum at which it becomes clear that 'the first decision is insensitive to [reasonable] variations in the future demand' (p. 1202).

The resulting investment commitment is made to invest in, construct, and begin operating the facility or facilities represented by that 'first decision' over the next time period. As the financial environment changes, as more data becomes available to provide new indications of the evolution of the socio-economic environment, as policies or regulations change, or whatever, 'reversible commitments' and the entirety of the originally proposed future investment stream may be reconsidered. 'Reversible commitments' may be understood to be those aspects of a commitment which may be altered. Indeed, depending upon the nature of the changing environment, any aspect directly related to physical construction may, of course, be altered. For example, new financial limitations may suggest a delay in the originally proposed date of completion to beyond the expected spring rains, although the original commitment to construction is maintained. Of course, the entire original commitment may be abandoned under sufficient new limitations.

If the reason for future alteration in commitments is presently unforeseeable, then the incurred sunk costs will be losses incurred as a function of the present uncertainties in such long-range ventures and should not be taken as cause for delaying all future similar endeavours which appear favourable under the same planning methodology. If the reason for the future alteration is sufficient to stop the construction of a partially completed dam – and leaving half of a dam may be worse than having no dam at all – then the indication is that the heuristic stopping criteria with respect to the horizon of the planner was grossly in error and it is the planner, and not necessarily the method that should then be replaced! In either case, the overriding present concern must be for changing the present system in some desired way, 'focusing on the present, enriching . . . life *as it* [*is*] *being lived* – with a mind to the future, of course, but with no obsession that the future itself [is] what [makes] planning worthwhile' (Ewing, 1972, p. 448).

Certainly future generations will not desire our pollution of the lakes, rivers, and streams and present life will as well be enriched by cleaning up the present mess. Within these well-defined objectives, it is only required to consider the present situation and derive a solution to it, circumventing the aforementioned uncertainties of long-range forecasts and their complications in defining 'optimal' short-range decisions. This simple decision-making exercise concerns a dynamic physical system within a static social value system and would, therefore, appear to be an ideal environment for the exploitation of the sophisticated mathematics of modern control theory as a planning tool. But the Loehman-Whinston (1970; 1973) – Forster-Logan (1973) exchange suggests that such a tool provides no additional insight beyond classical approaches and represents a confirmation of the von Neumann-Morgenstern (1944) observation noted in Chapter 3.

The problem posed initially by Loehman and Whinston (1970) is to develop the optimal form of control of pollution by an upstream firm so as to maximise the joint returns of that firm and another firm

curves on which these paths lie' (Nikaido, 1968, p. 207). Then, interpreting these optimal paths analogous to turnpikes in a highway transport system (cf: Samuelson, 1960) and adopting a finite-horizon perspective, this 'normative significance' may be incorporated into the arguments of this book in the following fashion. The further the planning horizon is extended into the future, the greater will be the influence of long-term optimality upon short-range decisions. Further, no matter how short the planning horizon (so long as it is strictly positive), there exists some influence by long-range optimality upon short-range optimality (cf: Intriligator, 1971, pp. 434-5). The moral is: no matter how myopic the planning of present society, it should incorporate hyperopic considerations.

Yet, the major problem plaguing mathematical planning of structural decisions is the availability and quality of data. Martens and Pindyck (1972) attempt 'to show that optimal control theory can be very applicable to long-range multi-sectoral planning if one is willing to work with a linear model and quadratic cost (or welfare) criteria' (p. 31). Through the formulation of a tracking problem, the resolution of conflicts in the sectoral allocation of investment resources by the Tunisian Central Planning Agency is confronted for the period 1972-81. But, the estimation of coefficients was conducted using a '136-dimensional 1968 input-output table' (p. 34); now, surely, not even the Tunisian economy is so static that the use of an optimal control-theoretic formulation with a four-year-old, finely disaggregated data set can be any more meaningful, in a realistic planning sense, than some less grandiose formulation using fresher, more stable, aggregated data for investment decision-making at the national level. Again, the von Neumann-Morgenstern observation is confirmed.

Townsend *et al.* (1973) analyse what may be termed an ideal centralised planning situation in the real world. Furthermore, whereas the above Tunisian example was concerned with the resolution of conflicts resulting from competition for scarce resources, Townsend *et al.* study planning by the British Gas Council whose charge is to coordinate the planning and operation of the regional Gas Boards. Now, the solution of a one-commodity system should be quite trivial if an entire economy can be planned and especially since the components cooperate rather than conflict. But, as so often is the case, simple systems admit efficient monitoring and short-term evaluation, thereby revealing the fallacy of the previous conjecture on the solution on a one-commodity system.

The first supplies of natural gas entered the United Kingdom from Algeria in 1964; the next two years saw discoveries of natural gas fields in the North Sea. Thus, in 1964, planners were provided a golden opportunity to design an optimal system from scratch. A 'specially constructed pipeline system' was designed to distribute the regassified import from Algeria; but, the North Sea discoveries were not planned into the system and the concern became 'to take full advantage of the plant which was already available' (p. 2). Furthermore, 'at the initial stage . . . insufficient accurate data on future demand levels and operation' was the fundamental problem in the physical realisation of the system (p. 4) and the relatively long life of pipelines necessitated that consideration be given to the long-range evolution of the system. As the system develops, a given pipeline may supply gas to different demand nodes; therefore, the 'final' system may be unnecessarily costly if storage, re-gassification facilities, and pipelines are poorly placed early in the investment process.

*Network Investment*

As a prelude to our specific treatment of road investment rubrics in the next two chapters, it is pertinent that we complete this section with a discussion of network structures.

> To develop general principles that might guide the community in deciding upon either improved or optimum transportation facilities, we are forced by the complexity of the problem to work with *simple* models. Expressed another way, we must study 'ideal-type' situations . . . We will hope . . . to provide some parables . . . each of which highlights a particular though uniquitous problem. From each of these we wish to draw a moral, a principle that ought not to be overlooked when a more complex situation is to be faced . . . This may still be good heuristics for the practical designer (Strotz, 1965, p. 128; emphasis added).

Alao (1975) echoes this concern for study and interpretation of 'ideal form' (pp. 1-2) and the 'needed insight' into the evolution of urban structure that it provides (pp. 20-1).

However, despite quite thorough mathematical developments on the part of both Strotz and Alao, neither reach any observations which add significantly to insight. In both, the problems have been abstracted to the point where only solutions of such an aggregate nature exist as to be virtually useless to those who must make decisions on such practical questions as congestion tolls and the extent of transport network development. More importantly, how do planners use 'the precise information about the quantity of land that should be devoted to transportation to cope with congestion at various points in the CBD' (Alao, 1975, p. 16) to optimally sequence public investment in a multimodal urban transport network of streets, limited-access roads, bus routes and rail lines. 'The price of complexity [may very well be] a restriction on insight' (Alao, 1975, p. 13); but the price of mathematical elegance in these two cases had been a failure to push forward the limits of understanding of the land use/transport system relationship. Alao has provided no new insights into this matter which were not available from Strotz a decade earlier and Beckmann, McGuire, and Winsten (1956) a decade before him. Indeed, operational understanding of the economic theory of networks has not substantially progressed beyond Ellet (1839), Dupuit (1844), Wellington (1887) and Hotelling (1938).

Concerning the topological design of networks, Dusonchet and El-Abiad (1973) present a 'new method' to tackle the problem of electrical transmission planning, a method fundamentally the same as Boyne's (1972) for gas distribution networks. This 'new method' is subsumed by Lin's (1975) heuristics in selecting from the large number of combinatorials a randomly-generated subset which lead to a subset of local optima which are then evaluated, one network being selected as optimal. The discussion evoked by Dusonchet and El-Abiad's presentation typifies the spectrum of thought surrounding the investment process in general and especially the nature of investment in locationally-fixed projects of a discrete nature.

Under any reasonably determined temporal sequencing of budget constraints, there are an enormous number of physically realisable combinations of projects and sequences thereof that could constitute an expansion strategy. Some of these sequences will probably be absurd practically but others will possess some possibility of satisfying at least the minimal or short-run criteria for implementation. On a theoretical basis, the study of the selection process can be separated into two camps: (i) the detailed analysis of a very few alternatives and (ii) the general analysis of a large number. In application, the first generally follows the second in the selection of commitments.

Consider the following situation: there is a natural gas company which serves different areas of a country. Assume that the company does not choose to employ liquified or synthetic natural gas; then, it must determine from which natural gas producer or producers and from which field to move gas and to which market. In the selection of the supply, reliability of supply, quality of the gas, and field and market prices must be considered. For example, the company may be able to contract for low-priced gas for ten years or for presently-high-priced gas of the same quality for twenty years from different locations. Managerial expertise in evaluating the financial future of the company, its expansionary policy, demand forecasts, supply price and location forecasts may bias any analysis in favour of the security of the longer-term contract but an engineering analysis may suggest that given the existing network the least-cost solution is to take locational advantage of the other source. Management decides that the regulatory authority will allow the additional investment to be added into the rate base. In an effort to abate depreciation's drain on the rate base, management opts for the 20 year contract with design specifications based on that horizon, thereby necessitating large initial investments to obtain economies of scale over the entire planning horizon.

Certainly, a real-world problem will not be so easily reduced in its complexity, but 'combining a straight-forward procedure with the most fundamental problem definition results in a practical approach to large planning problems' (Garver and Sigley, 1973, p. 1368). At the highest planning level, the problem of mapping strategy for a firm and of determining financial priorities does not find its answer within the framework of the exact details of a particular pipeline's construction.

A middle-level planning group must work within this 'grand' strategy to enable achievement of the goals: the introduction of simplification and generalities enables the forest, and not the trees, to be analysed, and it is in this phase that straight-forward methods are useful as a 'screening tool' for specific projects (Landgren and Anderson, 1973). At the level of implementation, such simplifications are no longer tolerable and the trees must be studied.

Therefore, in network planning, a universe of all combinatorial possibilities, which necessitates the use of theoretically sub-optimal constructions, is an 'unnecessarily super-abundant' representation of the actual decision problem, which includes only a relatively small set of logical realisations identifiable by an experienced planner (Brito, 1973). In a realistic, complex system exposed to future uncertainties, how would the analyst know if he had identified a 'good' or a 'bad' design? Lin (1975), following the philosophy of L. Zadeh (1965), defines a 'good' heuristic algorithm as one which 'produces solutions which are acceptable to the design engineers', since, in 'real-lfe situations, the problems are often so complex and unstructured as to defy analysis' (Lin, 1975, p. 42).

A similar argument may be applied to the method proposed by Wang, Snell, and Funk (1968) which involved a control-theoretic construction to determine which links of a grid pattern were to be selected to move vehicles from point demands to a CBD. Not surprisingly, a tree structure emerged. The problem did not deviate from the 'simple but important case . . . of a central city to be connected with outlying satellite cities' (Beckmann, 1967, p. 107) – and neither did their solution from his. Further, their solution gave no guidelines that would aid in the evaluation of a construction investment sequence nor was it at all clear that the method was better when applied to large transport systems than the more straightforward methods of, say, Steenbrink (1974).

One of the most fundamental characteristics of transport systems is that they evolve as a series of additions to an existing, though perhaps primitive system; as a system evolves, it influences and is influenced by changes in the political environment, land-use patterns, travel behaviour, and a myriad of other outside forces. Thus, there is a need to continually re-evaluate the investment process, both the nature of the additions and the timing of their becoming operational. Network investments possess long economic lives and, therefore, influence the development of the environment over a long operational life. For example, large investments in motorways to relieve congestion may have considerable influence on the realisability of a future public transport system due to its behavioural influences upon the public; it may also exercise influence over the expansion of industries which contribute the same type of pollution as the car, since its combination with the car renders pollution levels excessive and the industries may face curtailment.

Certainly, the future of the socio-economic environment must be considered even if it is not known with a great amount of certainty. But concommitant with the gains that may be had from providing flexibility to face uncertainty are the costs that may befall. For example, the additional right-of-way incorporated into the initial purchase of a new motorway in order to gain economies of scale with respect to potential future lane expansion biases any future re-evaluation of the subsequent investment process by providing 'cheaper' land for laying roadway than what may be had elsewhere. If the decision sequence is still optimal at that time, the future benefits from the present's sacrifice; if it is not still optimal, everybody loses. The result is binary and the initial expansionary programme should analyse the initial decision appropriately and not represent the future results by some nebulous construct of probabilities over many different situations.

As a final example of models, meaning, and manageability, Alper (1972) provides an interpretation relating to teachers and money that, he contends, 'would not be evident if instead of the mathematical model . . . we had a verbal model only' (p. 298). Unfortunately, Alper's solution reinforces rather than solves the problem faced in many professions of scarcity cycling with surplus of university graduates, with scarcity in one year leading to increased entering freshmen in the next and to a surplus of job applicants four years hence. Alper's handling of duality is too simplistic to be of insightful value: duals of mathematical programs do not interpret themselves but evolve from well-structured arguments (Balinski and Baumol, 1968; MacRae, 1969; Chan and Leininger, 1972). Therefore, that it 'goes without saying that neither of [Alper's] statements would have been readily obvious without control theory', (Alper, 1972, p. 301) is somewhat ludicrous in light of the fact that the interpretations are plausible mostly because they are readily obvious without reference to any mathematical constructs!

Then, we are left with the observation that good planning rests upon good data and good professional judgement.

## Conclusion

With this reference to recent state-of-the-art papers on optimal control approaches to investment planning we have provided a setting for road investment issues. Dommel (1975), in covering expansion planning for power networks, notes that further 'progress will probably be made through the use of *simplified models*' (p. 83; emphasis added). Gazis (1975), in dealing with the planning of transport networks, suggests that more effort should be directed toward such miniscule aspects as 'feedback in combining traffic assignment and origin-destination computations' (p. 76) and in 'investigating the value of a unit of investment in improving a network by selective improvements of a single arc' (p. 77). This concept of feedback between demand and the level of service is the foundation of Chan's (1972) airline route optimisation since profit 'to the carrier . . . is a function of the aircraft routes and passenger routings' (p. 508).

In summary, the formulation of any approach to investment analysis under uncertainty must explicitly incorporate the interaction of investment and the evolution of the controlling environment.

# 6 THEORETICAL ROAD NETWORK PLANNING WITH CERTAINTY

## Optimal Network Size

The social worth of a road network is that it provides capacity to move from place to place so as to reduce the operating cost of movement. Thus, the social cost of a regional or national road system (S) is composed of two elements. The first is the present value of the cost of capacity (K). These are costs of construction and maintenance with increase with the stock of roads (Q), as in Figure 6.1. The second part is the present value of the operating costs (T) associated with a particular stock of roads. This decreases as the stock of roads increases, diminishing circuity of routes and levels of congestion, as in Figure 6.1.

Figure 6.1.

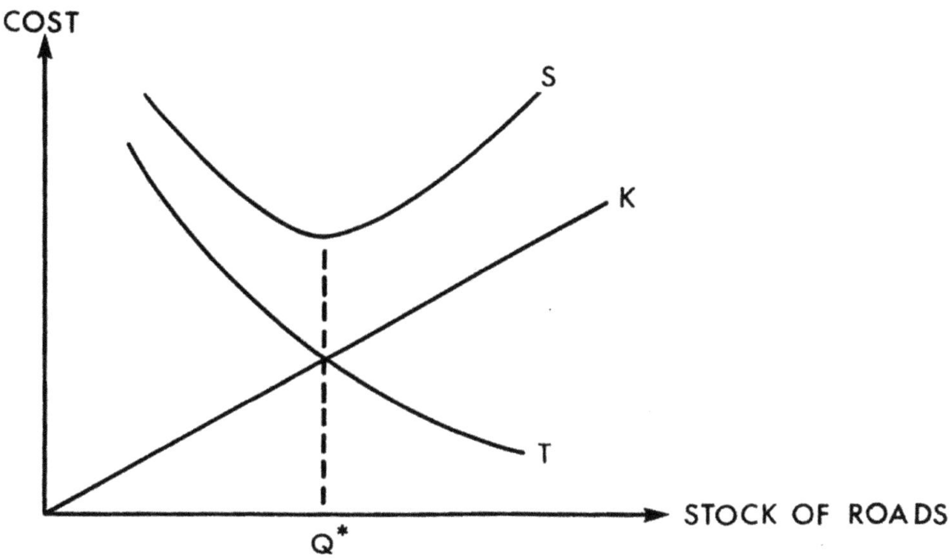


From this it is evident that the total cost of the road network will be U shaped and that there will be a minimum social cost stock of roads (Q*) lying somewhere between zero and infinity. If we knew the nature of the relevant function this level of stock could be established

DOI: 10.4324/9781003182993-6

by setting the derivative of the social cost function with respect to capacity equal to zero:

$$\frac{dS}{dQ} = 0 \tag{1}$$

Given the composition of the social cost function this minimum point would be found where

$$\frac{-\partial T}{\partial Q} = \frac{\partial K}{\partial Q} \tag{2}$$

at the point at which the marginal productivity of the road stock equals its marginal capital costs. Thus, it is worth continuing to increase the capacity of a regional or national network until the decline of operating costs just offsets the increase of capital costs. The stock of roads should be expanded while benefits measured in operating cost savings, exceed incremental capacity costs.

Road networks have been built up over an historical time-scale and current decisions are made in terms of the balance of costs and benefits of additions to a massive existing structure. As a starting point to review road investment decision rules when an assumption of perfect knowledge is made, we take the case of marginal increments.

**Independent Road Projects**

Suppose that additions to a road network can be taken to be independent of the existing system of roads and of each other in their outcomes. In effect we assume that the benefits generated by putting new or improved lengths of pavement in place accrue entirely to potential or existing users of that alignment without influencing the performance and use of other roads at all. If this were the case, a straightforward investment rule could be applied, given a measurement and value for net benefits (B), knowledge of construction or improvement costs (K) and a discount rate (r), trading future consumption off against current consumption. If the construction cost for a project were the same regardless of when it was executed and if the stream of net benefits were endless, then projects should be selected for action if the present value of the benefits they generate exceeds the cost of construction:

$$\int_t^\infty B(t)\, e^{rt} dt > K \tag{3}$$

This static formulation can be extended to treat the question of project timing. We can assume that the benefit rate is independent of project age and only depends on calendar time, i.e., that the project produces output at a constant rate from inception to infinity, which can in general be set to unity. In this case the net present value becomes:

$$Y(t) = \int_t^\infty B(t)\, e^{-rt} dt + rKe^{-rt} \tag{4}$$

To determine the best time to build the project we take the first derivative of this, which Marglin (1963) calls the marginal net present value of postponing construction, and set it equal to zero:

$$\frac{dY}{dt} = -B(t)\, e^{-rt} + rKe^{-rt} = 0 \tag{5}$$

which cancels to

$$B(t) = rK \tag{6}$$

The second order condition, ensuring that this is a maximum, is that:

$$\frac{d^2Y}{dt^2} = -r\frac{dY}{dt} - \frac{dB}{dt} e^{-rt} < 0 \tag{7}$$

Figure 6.2 Optimal Timing

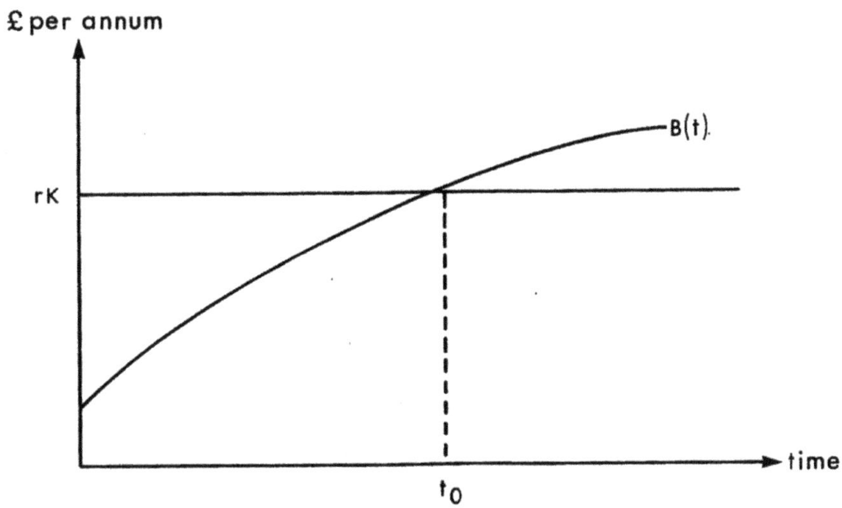

which holds provided that:

$$\frac{dB}{dt} > 0 \tag{8}$$

that is, provided the benefit rate is increasing when dY/dt = 0.

In the diagrammatic terms this is shown in Figure 6.2, where the optimal construction date ($t_0$) falls when the value of a project's output (Bt) catches up with its interest costs (rK).

If B(t) is monotonically increasing so that:

$$\frac{dB}{dt} > 0 \text{ for all t} \tag{9}$$

then the first order condition for a maximum yields a static rule for investment decision making, which may be repeated year after year to obtain the optimal starting date. Dividing equation (6) through by r we have:

$$\frac{B(t)}{r} = K \tag{10}$$

Thus, the best construction date $t_0$, is that at which the project's net present value is non-negative for the first time, where the benefit rate is assumed to continue indefinitely.

**Budgetary Interdependence**

Especially in the public sector, investment decisions are often bound by externally determined budgetary limit over discrete time intervals. This forces the selector into a simultaneous, comparative treatment of all projects. This budgetary interdependence can be treated as a programme for decision. Where the budget in any year t is $K_t$ and we can designate the net present value of the nth project in the tth period $Y_{nt}$, then the investment problem is to determine the value of the decision variable $\delta_{nt}$,

$$\sum_{n=t}^{N} \sum_{t=1}^{T} Y_{nt}\,\delta_{nt} \tag{11}$$

subject to

$$\sum_{n=1}^{N} K_{nt}\,\delta_{nt} \leqslant K_t \tag{12}$$

$$\delta_{nt} = 0 \text{ or } 1 \tag{13}$$

This constitutes an integer programme which solves the investment problem taking interdependence of projects resulting from an externally specified budgetary limit into account.

**Operational Interdependence**

We have already discussed the measurement of welfare associated with the use of transport facilities. Now we need to see how it is composed in the operation of a transport network, taking the effects on other parts of a network than the link proposed for building or improvement and effects on travel demand into account.

Let us start from the simplest configuration possible as shown in Figure 6.3. This network consists of an origin i and a destination j linked by a somewhat circuitous existing road m on which the generalised cost of a trip is $c_{ijm}$. We consider investing in a new alignment between the two points, labelled n. This will be more direct with better geometry and will thus have a lower cost of travel $c_{ijn} < c_{ijm}$. If the demand for travel is perfectly inelastic at $T^1_{ij}$ trips no matter the cost; if the volume travelling can never saturate the capacity of n and lead to congestion and if all travellers know and use the minimum cost route, then the change in consumer surplus brought about by building the new link would be: $T^1_{ij}$ $(c_{ijm} - c_{ijn})$, i.e., the saving in resource costs to existing users. If we allow for some responsiveness of demand to reduced travel costs, so that a greater demand $T^2_{ij} > T^1_{ij}$, will result from the lower cost $c_{ijm}$, and if we can assume the demand curve for transport to be linear in the vicinity of the cost change, then the consumer surplus increase associated with this link will be:

Figure 6.3

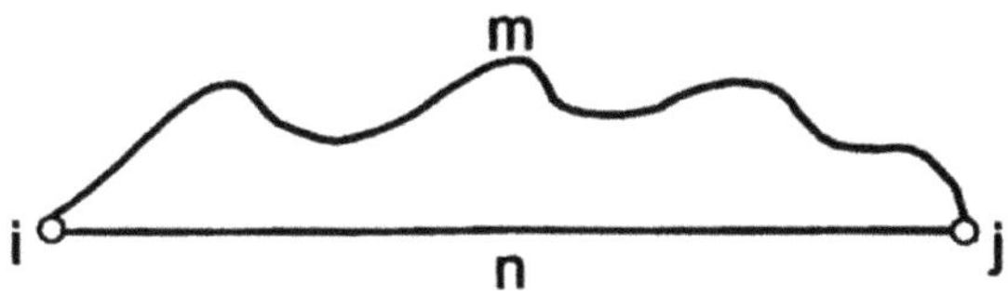


$\frac{1}{2}$ $(T^1_{ij} + T^2_{ij})$ $(c_{ijm} - c_{ijn})$. Road system costs are governed by relationships between speed and flow, reflecting congestion effects. As flow increases beyond a certain point, so velocity falls. If we allow cost to increase with volume and assume that travellers are indifferent between routes of equal travel costs, then we can envisage an equilibrium where the costs on both routes will be equal and some travellers will be found on both links: $\overline{c_{ij}} = c_{ijm} = c_{ijn}$ and $T^2{}_{ijm}$ and $T^2{}_{ijn}$ both occur.

Under such circumstances the welfare change would have to be measured from both links: $\frac{1}{2}(T^1_{ij} + [T^2_{ijm} + T^2_{ijn}])(c_{ijm} - \bar{c}_{ij})$, where $\bar{c}_{ij}$ is the equilibrium generalised cost after the adjustment in total volume and in choice between routes m and n has settled down.

It is evident that an evaluation of one project must consider the effects on competing routes and we can indicate this generally by writing the benefits expression as the sum over all route options between the origin and destination:

$$\frac{1}{2}\sum_{i,j}(T^1_{ij} + T^2_{ij})(c^1_{ij} + c^2_{ij}),$$

where the superscripts refer to demands and costs with and without the project in question.

In reality we are usually faced with networks of many more than two nodes and we have to treat of complementarity as well as competition among routes. Roads m and n would usually be parts of paths connecting a great many other origins and destinations. Improvements between i and j will increase demand from further afield and thus affect costs on other links. For example, if the attractiveness of j as a centre for shopping is enhanced by building road n, then traffic may well be diverted from other stores, reducing demand and travel costs on routes serving them, and increasing volume and costs on links leading into j via i. Thus, in order to capture the full effects of a network change, such as the building of n, changes in demand and generalised cost must be traced over all elements of the network affected. In dealing with urban roads especially, this has led some practitioners to do the calculations for all possible origins and destinations under the assumption that the intricacies of urban traffic and activities are so meshed that the whole system must reverberate to a change in any of its members. So the benefit of a scheme is estimated over all z zones in a town with:

$$\frac{1}{2}\sum_{i=1}^{z}\sum_{j=1}^{z}(T^1_{ij} + T^2_{ij})(c^1_{ij} - c^2_{ij}) \qquad (14)$$

or more generally

$$\sum_{i}\sum_{j}\int_{c^1_{ij}}^{c^2_{ij}} T_{ij}(c_{ij})\, dc \qquad (15)$$

for all relevant i, j. The relevant field of network modes is given by the limit where network perturbations from a particular source wane to insignificance.

Clearly such a specification of the optimand of the investment programming problem makes it a much more cumbersome matter. Much of the modelling of land use and transport systems undertaken in the last twenty years has been directed at the symbiosis of accessibility and activity locations and the forecasting of outcomes of transport improvement, with little demonstrable success. A recent attempt to integrate network development and land development by Putnam (1975) failed to converge to a stable trajectory. Thus it seems that we are some way off from the analytical incorporation of locational shifts in the transport benefit measurement.

## Numerical Models of Optimal Network Improvement

If a suitable and readily measurable social objective of road planning can be identified the evaluation process can be rearranged to become an optimal design procedure. Methods of mathematical optimisation can be applied to compute the location and capacity of road improvements which will maximise some measure of social welfare, if some sacrifices of reality are made. Mathematical programming methods deal with discrete space and so the transport system must be represented as a graph of nodes and links. The solution comes out in terms of whether two nodes should be directly connected and what the capacity of that link should be. In the face of ignorance of the process of interaction between land use and accessibility involved and in the interest of computational tractability, a frequent recourse in such fomulations is to an assumption of given total demand for travel and a fixed matrix of trips between places to service. This translates the maximisation of the consumer surplus measure of welfare as an optimand into the minimisation of the social costs of travel, with which we commenced this chapter.

To grasp the nature of the necessary simplifications involved we can trace the evolution of these numerical methods of optimisation as applied for road investment programming.

### *Linear Programming*

Numerical procedures to find the extreme value of a linear function within a domain of feasibility defined by linear constraints were devised for the solution of military logistic problems during World War II (Dantzig, 1963). The achievement of a quantum leap in the speed with which calculations could be done by computer enabled their widespread commercial and administrative application. The methods were extended to a wide range of practical planning problems chiefly in the field of industrial engineering. One fairly obvious extension from the original logistic problem was to the design of networks, be they for communication or transport. Quandt (1960) augmented the fundamental Transportation Problem of Linear Programming so as to solve not only for optimal shipments between supply and demand points but also for the best capacity on the network links joining them. Formally the objective is to find the flows $T_{ij}$ which minimise the total cost of shipments of a homogeneous product where unit cost of shipment is $c_{ij}$:

$$\text{MINIMISE} \sum_i \sum_j T_{ij} c_{ij} \tag{16}$$

whilst ensuring that the requirements ($D_j$) and availabilities ($S_i$) at various points are satisfied by nonnegative flows:

$$\sum_i T_{ij} = D_j \tag{17}$$

$$\sum_j T_{ij} = S_i \tag{18}$$

$$T_{ij} \geqslant 0 \tag{19}$$

The links between places are construed to be capacitated ($U_{ij}$) and simultaneously with solving for flows, the programme must determine the best increments to existing capacity ($\Delta U_{ij}$), at construction cost $K_{ij}$, which can be built from a limited budget for transport investment K, so as to reduce the objective by the largest amount:

$$T_{ij} - \Delta U_{ij} \leqslant U_{ij} \tag{20}$$

$$\sum_i \sum_j K_{ij} \Delta U_{ij} \leqslant K \tag{21}$$

The maximising dual of this programme yields an imputed rate of return (r) on the marginal increment of road capacity i.e., the decrease in the primal optimand resulting from an additional pound spent on expanding road capacity:

$$\text{MAXIMISE} \sum_j v_j D_j - \sum_j u_i S_i - \sum_i \sum_j w_{ij} U_{ij} - rK \tag{22}$$

subject to

$$v_j - u_i - w_{ij} \leqslant c_{ij} \tag{23}$$

clean air, other uses of land, safety and an extension of the life of oil supplies, may change radically with political fashion. If anything is certain it is that the budget for road expenditure will be subject to radical amendment from current expectations. Funds committed to road building have always been an easy target in times calling for austerity and cut-backs in government spending. We have recently witnessed the virtual disappearance of the road programme in the UK as an austerity measure with minimal political cost compared to the cutting of other expenditures. In such an instance, all previously laid plans count for nought.

At this juncture we have provided the groundwork upon which to base a review of the methods proposed for the incorporation of uncertainty in road planning rubrics.

# 7 ROAD PLANNING UNDER UNCERTAINTY

## Maximisation of Expected Utility

As descriptive theorising on decision-making unfolded through the 1950s and 1960s, accompanied by the elaboration of operational optimising techniques, so general analytic approaches to uncertainty were translated into transport investment terms. Reutlinger (1970), as a staff member of the World Bank, presented some views and prospects for appraising projects in terms of the means and variances of probability distributions of rates of return approximated by a simulation of the relevant variable values and discussed the appropriate attitude toward risk. In a companion volume, Pouligren (1970) reported the application of such probability appraisal by simulation to two major transport projects with which the Bank was involved. One was the Port of Mogadisço in Somalia and the other the Tanzam Highway. In the latter case the usual level of uncertainty about such investments was compounded by the prospect of the Tanzam Railway being constructed with Chinese assistance and its likely effect on road traffic.

This approach can be characterised by contrast with the extreme simplification of a 'certainty' appraisal. Certainty implies the ability to determine an outcome precisely. In evaluating two projects A and B, the value of the returns to each would be known exactly as in Figure 7.1 and the best scheme, A could be selected unambiguously.

Supposing we are faced with a risky situation where we can only put odds, of varying degrees of precision, on the likelihood of a particular outcome. This would give a spread of possible returns to each project with varying probabilities as in Figure 7.2.

A certain ambiguity over which is the better scheme now arises because the distributions of their associated returns overlap. The decision must revolve around the trade-off between a mean expectation of high returns or low costs and the greater security of a low variance about the expected value. In the example, project A has the lower mean social cost but also a higher variance. The lower benefits associated with B are a much surer thing. The question then becomes a matter of the degree of risk aversion of the decision maker – whether he is neutral to risk, a gambler or averse to taking chances. This can be represented as the trade-off function between the mean $\mu$ and variance $\sigma^2$ of the outcomes as in Figure 7.3.


DOI: 10.4324/9781003182993-7

Figure 7.1

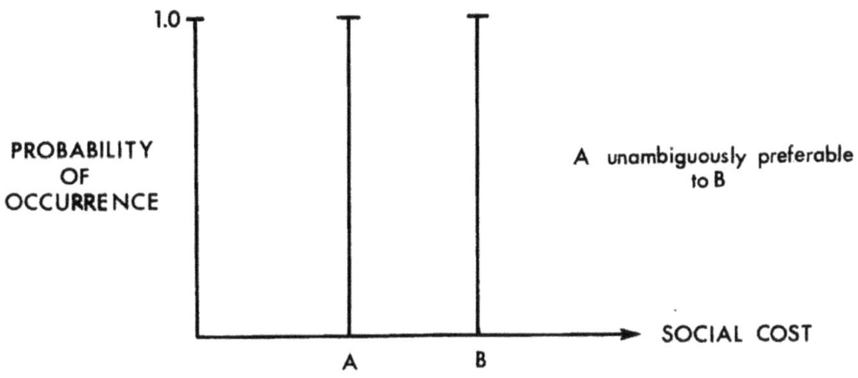


Figure 7.2

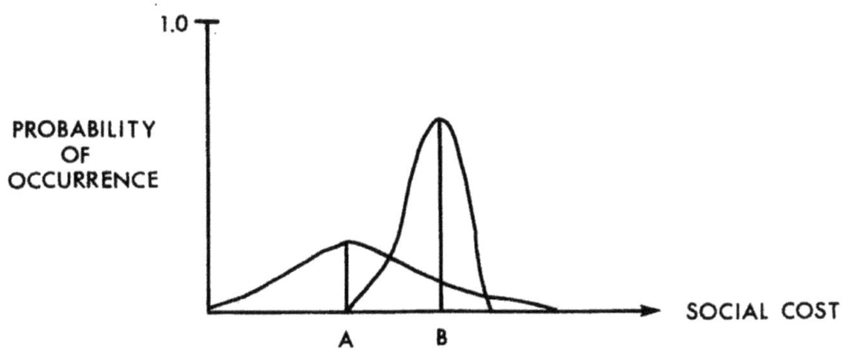

Figure 7.3

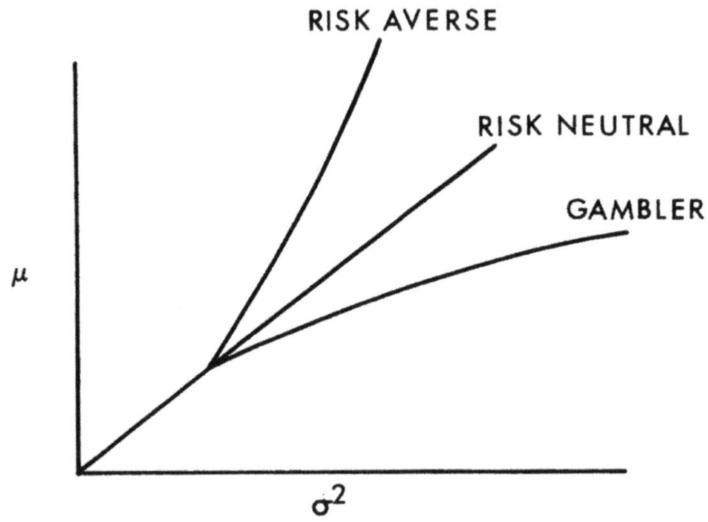


Someone with an aversion to risk might select project B despite its higher mean social cost because its outcome is much surer than that of project A, which has a distinct chance of bombing out and doing much worse than B.

Cole and Roberts' contributions to the Meyer and Strazheim volume (1971) offer a synthesis of these developments. Cole presents a procedure for the treatment of uncertainty at the project design stage of planning where it is legitimate to construct subjective probability distributions of the performance of a number of discrete designs for a facility. Under an assumption of constant marginal utility of income, the design with the lowest expected total cost is then chosen.

Roberts presents a programming formulation for the maximisation of expected utility, measured by some cardinal index. The arguments of this utility function and the benefits of possible projects are stochastic and capable of representation by subjective probability distributions. Roberts first claims that in most transport planning it is reasonable to assume a linear utility function. This makes it possible to ignore higher moments of probability distributions of outcomes and merely solve straight-forward mathematical programmes with expected values in place of point estimates of costs and benefits.

His arguments for the linearity of utility involve the small size of the transport programme in national budgets, implying that the variance of outcomes in this sector will be relatively unimportant. He holds, in addition, that the portfolio of transport projects will have a lower variance than that for its individual elements. As long as projects generate fairly immediate benefits, then he expects public officials to be willing to assume distant risks of loss with equanimity.

Roberts concedes that there are circumstances where there is an aversion to risk in terms of the trade-off between the mean and variance of project outcomes. If decision makers can specify this trade-off as the coefficient, a, of a negative exponential utility function, then this can be incorporated in the decision calculus. Suppose the present value of known variances of project outcomes $\bar{\sigma}_i$, their means $\bar{\mu}_i$, and covariances with other project outcomes $\bar{\sigma}_{ij}$, discounted at a given rate, can be determined. The selection of projects which maximises expected utility within some budget $K_t$ can be achieved by maximising:

$$\sum_j \sigma_j \mu_j - (a/2) \sum_i \sum_j \delta_i \bar{\sigma}_{ij} \delta_j \tag{1}$$

$$\text{s.t.} \sum_j K_{jt} \delta_j \leqslant K_t \tag{2}$$

$$0 \leqslant \delta_j \leqslant 1 \tag{3}$$

where $\delta_j$ is the fraction of the project j to be built or, in the discrete case, zero or one. The assumption is that whereas benefits are stochastic, construction costs $K_{jt}$ are known with certainty, so that budget limits are precisely met in the continuous case. If the costs are also stochastic then chance constrained programming can be employed. In this only projects for which the expectation of meeting the budget constraint exceeds a given probability ($\alpha$) are selected. The financial limitation then becomes:

$$P\{K_t - \sum_j K_{jt} \delta_j\} \geqslant \alpha \tag{4}$$

Roberts suggests that this may be an unnecessary frill since 'financial outlays can often be forecast rather accurately relative to the probability distribution of net benefits' and 'if the distribution of expenditures have finite second moments, recourse to the central limit theorem will assure that planned budget allocations will be reasonably closely approximated by necessary outlays'.

The first contention puts too much faith in the continuity of incremental gradualism in national and local budgeting and flies in the face of quite dramatic changes in the allotment of national funds. In governmental circles the ideas of zero based budgeting and the continual monitoring of all programmes and expenditures with a view to the routine elimination or curtailment of defunct items are gaining rapidly in popularity, especially in response to voter antipathy to high rates of taxation supporting massive expenditures. The California vote on Proposition 13 is symptomatic and has riveted government attention on this issue far and wide. In some state governments, the principle of continuous review is enshrined institutionally, as in Florida's Sunset Laws. These trends and events should sound a tocsin against excessive confidence in the status quo.

The appeal to the central limit theorem is based on the expectation that over and underestimates over many projects, where cost probability distributions have limited ranges, will cancel out. There is an implied assumption of symmetry about expected values. Since cost overruns are more characteristic of reality, this seems dubious.

### Demand and Performance Interdependence

The programme presented above treats of dependencies among projects resulting from capital budgeting but not demand and performance relations, the feedback effects of demand on the performance of the network and the influence of the transport system on location decisions over the drawn out sequence of actions of which network building consists. Pecknold and Neumann (1972) attempted to overcome these limitations by allowing for the interdependence of network structure, land use and demand for movement. If we seek the best evolutionary sequence for a network allowing for such connections, the probability distribution of demands at any stage are dependent on actions and states at previous stages or over sequences of stages. Clearly the decision space which must be searched for some best sequence of actions, explodes to enormous proportions for anything but piddling problems. The computation of the network flow equilibrium associated with every action-state couple is very expensive. Pecknold and Neumann sought to limit the extent of search and the expense by use of a Bayesian decision tree schema with probabilities assigned to the various demand levels. The new demands predicted to emerge after the realisation of particular investments are used to recompute demand probabilities. The method assumes that the initial prior distribution of variables describing states of the world is known exactly along with the function describing the demand for transport which remains constant in shape throughout time. When examined closely, the powers of foresight required by this technique are formidable and quite unrealistic.

**Decisions with Limited Foresight**

The formulations reviewed thus far require knowledge sufficient to compute the probability of various configurations of demand and performance associated with particular courses of action. Very often this is a ludicrous requirement and it would be better to plead ignorance. Pearman (1976) offers a procedure for circumstances lying between precise knowledge of probability density functions of outcomes and the admission of complete ignorance. He suggests that, frequently, sufficient information may exist to allow of a ranking of the likelihood of future states of nature. If the payoffs, in our case the present values of the social costs of transport, associated with certain actions, given the state of the world, are known, then it is possible to calculate the minimum and maximum expected payoffs for any single network investment strategy i consistent with known payoffs ($c_{ij}$) and a subjective, ordinal ranking of the probabilities of states of nature j:

$$\text{for any i, minimise } E(S) = \sum_j p_j c_j \tag{5}$$

$$\text{s.t. } \Sigma p = 1, \quad p_j - p_j + 1 \geqslant 0, p_j \geqslant 0 \tag{6}$$

Should the central value of the payoff be insufficient evidence, the analysis may be extended to calculate the minimum and maximum values of variance

$$\text{var}(S) = \sum_j p_j c_j^2 - (\sum_j p_j c_j^2) \tag{7}$$

Such measures would only apply to a set of independent investment projects and they do depend on precise knowledge of the payoffs of action-state couples.

**Decision Making under Ignorance**

There are several general ploys available when we cannot judge the likelihood of future states of the world on any cardinal or ordinal scale but can estimate the payoff associated with each possible combination of action and state. Suppose we have several different network structures labeled $N_1 \ldots N_n$ as investment options and that we can identify m possible future states of the world $S_1 \ldots S_m$. The social cost of transport associated with network configuration $N_i$ and state of the world

$S_j$ is known to be $c_{ij}$. For five network structures serving four cities under three different states of the world we can construct a table of the values of the social cost of transport (a matrix $| c_{ij} |$) associated with the different networks under possible futures as in Figure 7.4

The Bayes or Laplace criterion assigns equal probabilities, p, to each of the possible outcomes and uses these to compute the expected value of each action, so as to select the best. Which in this case, where payoff is in the negative terms of costs, would be the minimum:

$$\text{minimise } E\pi_i = \sum_j p\, c_{ij} \tag{8}$$

The results obtained do obviously depend on the number of equally likely outcomes identified at the outset.

The Minimax approach goes to the conservative extreme of determining the worst possible outcome for each option available. The action with the lowest maximum value is then selected, thus minimising the worst possible costs. This is the strategy of purest cowardice, which in some circumstances may be the most rational.

The Minimax strategy goes for the long shot, appealing to more heroic, and potentially dangerous, instincts by selecting the project with the largest maximum payoff, i.e., the lowest minimum cost.

To balance out these two extreme tendencies Hurwicz (1951) suggested a weighted average of maximum and minimum values should be used. For each network $N_i$, we let $c'_i$ and $c''_i$ be the maximum and minimum values of transport cost. We define a pessimism-optimism index $\alpha$ such that $0 \leqslant \alpha \leqslant 1$. For each network $N_i$ we define its $\alpha$ value as $\alpha c'_i + (1 - \alpha)c''_i$ and choose the option with the highest $\alpha$ value. The criterion generalises the Minimax and Maximax criteria. By setting $\alpha = 1$ Minimax is invoked and $\alpha = 0$ employs the Maximax principle. Intermediate values mix the two extremes. This implies an arbitrary judgement of weights and like its limiting cases it discards information about intermediate values in favour of the extremes.

Savage (1951) applied the Minimax criterion to a measure of the regret or opportunity cost associated with a decision turning out wrong. Regret is measured as the difference in payoff between pairs of action-state couples. For each state of the world $S_j$ let $c'_j$ be the lowest cost attainable under that state i.e., $c'_j = \min_i c_{ij}$. With this definition we can construct a table of the regret associated with action-state combinations with entries $r_{ij} = c'_j - c_{ij}$. This matrix registers what is foregone for any action-state representation by the action chosen. If the action chosen turns out to be correct then the regret is zero. The Minimax criterion is then applied to the regret matrix, selecting the choice

Figure 7.4

STATES OF THE WORLD

| NETWORKS | $S_1$ | $S_2$ | $S_3$ | | Regret $S_1$ | Regret $S_2$ | Regret $S_3$ |
|---|---|---|---|---|---|---|---|
| $N_1$ | 3 | 11 | 3 | | 2 | 5 | 0 |
| $N_2$ | 6 | 6 | 6 | | 5 | 0 | 3 |
| $N_3$ | 1 | 11 | 11 | $= \lvert c_{ij} \rvert$ | 0 | 5 | 8 |
| $N_4$ | 2 | 9 | 9 | | 1 | 3 | 6 |
| $N_5$ | 5 | 10 | 7 | | 4 | 4 | 4 |

$= \lvert r_{ij} \rvert = \lvert c'_j - c_{ij} \rvert$

SOCIAL COST OF TRANSPORT REGRET

which has the minimum value of its maximal regret. This obviously again focuses on extreme payoffs.

These criteria can be extended to incorporate several objectives, if the rate of exchange between them is known. If only a ranking of preferences is meaningful, it is possible to use an 'outranking' relationship to delineate a 'kernel' of actions no one of which is clearly preferable. Closer attention to these may then reduce doubt and incomparability of objectives (Roy, 1971).

Milnor (1954) showed that the application of these different ignorance criteria can yield completely different best choices among possible actions. The values in Figure 7.4 illustrate this. Assigning equal probabilities to each possible state on the basis of insufficient knowledge and computing the expected value of each proposed network, we can employ Bayes strategy and select the network which minimises expected cost i.e., network $N_1$. Remembering that our objective is to minimise the social cost of transport, we employ the Minimax strategy by selecting the network with the lowest maximum value of cost, namely $N_2$. The Maximax strategy in this case selects $N_3$, the network with the lowest minimum cost. If we apply a weight of ½ to the worst and best outcomes to reflect our degree of pessimism or optimism, Hurwicz's strategy would prescribe $N_4$. Constructing a table of regret and applying the Minimax criterion to this, à la Savage, selects network $N_5$.

**Flexibility as an Objective**

The above illustration of the divergence of choice points up the power of subjective differences in determining what is best. Forcing the circumstances of decision into a simple extremum seeking mould conceals but cannot satisfy the real requirement for expression of attitudes and allowance for their play in the selection of appropriate actions. The overrigid specification of a long term goal makes things tidy but possibly silly.

It is evident that the attitudes reflected in these ignorance criteria yield completely different best choices. Since attitudes may change and can radically alter the definition of 'best' through time then, clearly, flexibility is a desirable quality to build into any system. Decisions are most frequently made in a continuing sequence. This is certainly so in the case of road investment. In building up a road network over time the potential exists for modifying the geographical configuration which is emerging as new information comes to light and new attitudes form. It is not necessary to commit the system inevitably to a particular future shape. Distinguishing between a plan and a decision, it is only a decision and action which represents an irreversible commitment of resources. A plan is a set of prospective dated decisions which may be reversed or revised prior to their due date, in the light of new knowledge or changed circumstances.

In recognition of these realities Gupta and Rosenhead (1968) put forward 'robustness' as a decision criterion for sequential decision making under conditions of ignorance. In statistics 'robust' implies insensitivity to assumptions. A decision whose rude health will enable the system to perform well no matter what befalls, is robust. Robustness is the quality of useful flexibility.

The criterion is intended for use in planning facilities where probabilities cannot be assigned to the future behaviour of users, competitors, ancillaries and government and their attitudes and priorities. If we cannot in all honesty put probabilities on outcomes and can only specify a finite number of possible futures between some limiting extremes, but expect an unfolding through time of what the desired future might consist of, we would wish to be able to adapt the uncommitted stages of a plan. If the possibility of such modification is not built into the planning framework so that it influences early decisions taken, then there may be insufficient play in the structure to respond to improved perception of needs, goals and prospects. To quote the authors of the notion:

> All decisions limit the future by committing the present. A plan whose initial decisions limit the future as little as possible has an evolutionary advantage in an uncertain world (Rosenhead, Elton and Gupta, 1972, p. 418).

How can this quality be measured for operational use? The matrix in Figure 7.4 represents 15 maps, a set M of combinations of actions and states of the world. One of these will be realised. Any decision to build one link $\ell$ of the network will restrict the attainable futures to subset $M_\ell$ of all possible maps. Suppose we are unsure now in our attitude to road planning, whether we should be heroic or cowardly or Bayesian. In this case, we would wish to take steps now which will leave the adoption of any of the strategies open to use for as long as possible. The robustness of deciding to build link $\ell$ now is given by $M_\ell/M$. This ratio is unity for the diagonal link indicated in the diagram. So we maximise robustness by selecting this link as the first step in a sequence of decisions, switching the emphasis from a final plan to continuous planning.

In general in the road planning contest we are faced with n links of a network which can be built or improved. At any time a decision in favour of building link $\ell$ ($\delta_\ell = 1$) can only be made for a limited number of links. We conceive of the future as a set of possible action-state maps M, of which one will come about in the long run. Any present decision $\delta_\ell = 1$ will reduce the futures available to a subset $M_\ell$ of M. For some subset $\bar{M}$ of M which are presently judged 'good' final outcomes on the variety of criteria involved, a subset $\bar{M}_\ell$ will still be potential after the present decision $\delta_\ell$. The useful flexibility of link $\ell$ then is given by:

$$f_\ell = \frac{(\bar{M}_\ell)}{(\bar{M})} \tag{9}$$

Present steps in a sequence of decision should be chosen so as to maximise this quantity.

Successful design has always implicitly incorporated robustness in long-lived works and plant. Structures were erected to meet dimly foretellable contingencies at a minimum cost and yet to serve immediate needs quite well. The evolutionary process favours adaptable things of high tensile strength. Platonic specialisation and optimality carries the seed of its own failure. Political traditions and social order have succeeded best where they fostered generality and avoided the fragility of overspecialisation to ensure long term stability. Flexibility and resilience of structure are attained by having rugged components which can easily adapt to new circumstances.

In effect the Rosenhead and Gupta criterion formalises a sound, conservative tendency to discount the arrogance of oversimplicity, displaying a more truthful humility with respect to what is known and knowable.

**Robust Road Planning**

The idea of robustness as a decision criterion could well have been conceived with road planning in mind. In fact it was not and Gupta and Rosenhead (1968) illustrated the notion with an application to industrial plant location. However, the criterion was applied soon after its conception to local government finance and planning where road investment was a subset of the total problem (Friend and Jessop, 1969). The quantitative measure of robustness is presented in a road network context from the outset here because it is so obviously suited to that purpose. The road planning problem is long-termed, sequential, faced with many conflicting objectives and subject to great uncertainty as to outcomes and preferences. It is the epitome of circumstances where constant reappraisal and readiness to adapt ought to prevail in planning strategy. The uncertainty formulation reviewed in the earlier part of this chapter treated the realities of doubt as an irritating departure from an ideal determinate world. The strategy suggested by robustness is not a matter of adjusting the conventional optimality criterion to uncertainty but of recognising the realities and complexities of the decision problem in the first place.

# 8 A CASE STUDY OF ROAD PLANNING IN AN UNDERDEVELOPED COUNTRY

**Doubt and Development**

In very few fields of planning is the degree of uncertainty greater than that experienced in transport investment planning in developing nations. Quite typically, these countries are faced with extremely scarce resources and must determine investment priorities in all sectors of the economy in order to tackle simultaneously the numerous, seemingly insurmountable, obstacles to development. There are two principle consequences of this decision-making environment. First, there is an implicit but significant competition for the available investment funds among the various sectors of the economy. The resulting opportunity costs of capital are of vital importance since the consequences of investment misallocation are potentially disastrous. Secondly, the decision-making environment stresses the importance of a comprehensive analysis and comparison of all investment options. In other circumstances this might be feasible. Unfortunately, the inadequacy and unreliability of the data sources in these countries precludes ready analysis. Even within the transport sector itself the quality and coverage of existing data prohibits a far reaching analysis of the direct and indirect benefits of most investment projects. More often than not the estimation of benefits is limited to direct user cost savings and indirect benefits are assumed to be either negligible or impossible to estimate.

In addition to the uncertainties which exist because of the shortage of investment capital, the massive demand for these funds from every sector of the economy and the difficulty of cross sectoral comparison, the very nature of the transport systems and economies of developing countries also introduces sources of doubt about the future. With rudimentary economic structure, frequently based on the export of a few primary agricultural or mineral products, underdeveloped nations are often prone to cataclysmic changes of fortune. Their terms of trade are dictated by the world economy and, thus, beyond the control of the individual state. In this climate, accurate forecasts of the demand for transport are well-nigh impossible, particularly for the long planning horizons required for transport investment appraisal. In addition, with poorly developed transport networks, investment programmes can, and are often intended to, precipitate massive changes in the geographical

DOI: 10.4324/9781003182993-8

structure of the economy. Planning is not aimed at a balanced evolution of the spatial status quo but at the radical rearrangement of activities in a more efficient array. A planning strategy recognising the implication of these sources of uncertainty is clearly called for.

The purpose of this chapter is to demonstrate the modification of a conventional interdependent road network investment model to reflect at least some of the degrees of uncertainty present in developing nations. First, an operational road network investment model is discussed and its use under conditions of complete certainty is described. This is followed by the presentation of an alternative view of the problem, one which incorporates some elements of uncertainty. Various decision-making strategies for this revised problem are examined and their use is illustrated in the context of a case study of a road investment planning problem for a province of Indonesia.

## A Road Network Investment Model

For operational purposes we need to elaborate on the road network investment programming schema presented in Chapter 6. A network investment model must account for the following elements. The network consists of nodes, which are centres of production and consumption, dummy nodes, which are network intersections and serve neither as origins nor destinations of traffic, and links, which correspond to individual transport routes. Associated with each node in the network is a level of supply and/or demand for each of a set of commodities which flow over network. Associated with each network link is a unit cost of travel and a link capacity. Commodities move over the network between the origins and destinations incurring transport costs. Investment in transport network infrastructure can take either of two forms. First, it is possible to upgrade or rehabilitate any existing transport link, reducing the unit cost of movement, increasing its capacity, or some combination of cost reduction and capacity expansion. Secondly, entirely new links can be added to the network. Any number of distinct investment projects can be associated with any network link. For example, it is possible that a network link could be improved from two lanes to four lanes or to the standard of a divided four lane facility. Different surface types and lane widths might also represent mutually exclusive investment alternatives. The network investment problem is to select and time phase the construction of the subset of investment projects that minimises the total vehicle operating cost over the network, subject to an exogenously supplied upper bound on the level of investment.

Suppose then that the transport network consisting of a set of N nodes is given and connected by a set of links over which G commodities are routed between their points of supply and demand. Associated with each link in the network is an investment project for which the existing generalised cost of travel over the link $c_{ij}$ and capacity $U_{ij}$ can be improved so that either the unit cost of the travel declines to $c'_{ij}$, the link capacity is extended to $U'_{ij}$, or else some combination of cost reduction and capacity extension occurs. Let $T_{ijg}$ be the flow and $c_{ijg}$ the cost per unit of flow of commodity g over the link prior to improvement and $T'_{ijg}$ and $c'_{ijg}$ the flow and unit cost upon completion of the project. (See Figure 8.1). If the traffic volume

$$T_{ij} = \sum_{g=1}^{G} T_{ijg}$$

is less than the upper capacity of the unimproved route $U_{ij}$, then total vehicle operating cost is calculated using the unit cost of travel over the unimproved link, so that,

$$Z_{ij} = c_{ij} T_{ij}$$

where $Z_{ij}$ is the total vehicle operating cost on link ij. If the link is improved to the higher standard then total vehicle operating cost is calculated using the unit cost of travel over the unimproved road, so

$$Z'_{ij} = c'_{ij} T_{ij}$$

Flow over the improved and unimproved links represent mutually exclusive alternatives. Either the link is improved and the lower costs and/or higher capacity apply or else the link remains unimproved and the existing cost and capacity apply.

All nodes in the network are subject to node-flow equilibrium constraints which ensure that their supply and demand requirements for the G commodities are all met, that is,

$$\sum_{i=1}^{N} (T_{ijg} + T'_{ijg}) - \sum_{i=1}^{N} (T_{jig} + T'_{jig}) = D_{jg} - S_{jg} \qquad \begin{matrix} j = 1, \ldots, N \\ g = 1, \ldots, G \end{matrix} \qquad (1)$$

This requires that the total inflow of any commodity g into any node j plus its supply, $S_{jg}$, equals the total outflow plus the demand, $D_{jg}$. Since local supply is assumed to be consumed locally (if necessary), any node will have either an excess demand or excess supply but not both. Dummy nodes are constrained so that total inflow and outflow are

Figure 8.1: Vehicle Operating Costs at Different Traffic Volumes over the Improved and Unimproved Roads

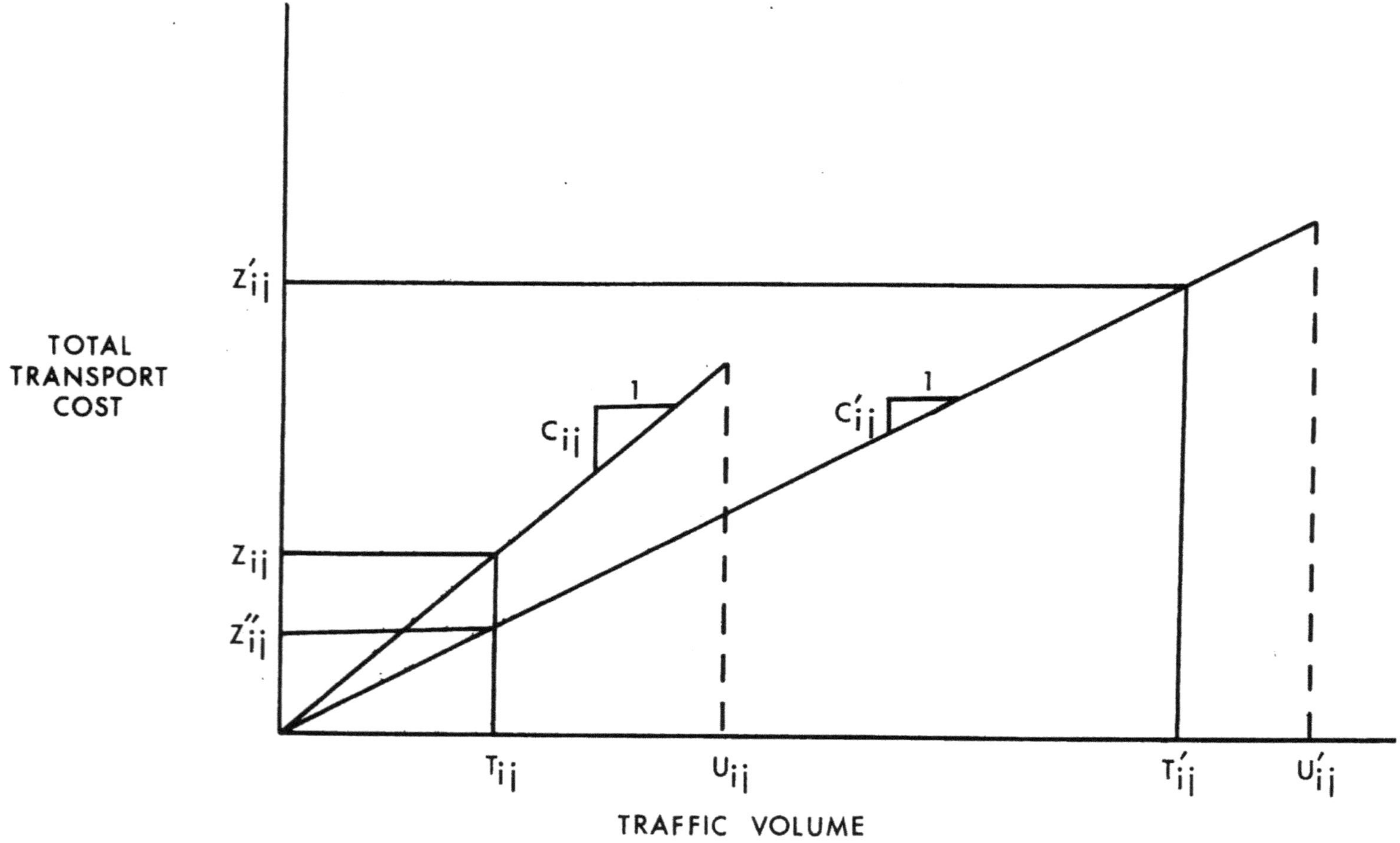

equal since there is no local supply or demand.

All commodity flows over the network are constrained to the capacities of the unimproved or improved roads, so that

$$\sum_{g=1}^{G} (T_{ijg} + T_{jig}) - U_{ij}(1 - \lambda_{ij}) \leqslant 0 \qquad ij\epsilon L \tag{2}$$

$$\sum_{g=1}^{G} (T'_{ijg} + T'_{jig}) - \lambda_{ij}\, U'_{ij} \leqslant 0 \qquad ij\epsilon L^* \tag{3}$$

where the set L consists of all unimproved links and the set L* contains all links for which an investment project is defined. The binary link investment variable $\lambda_{ij}$ takes on the value one if link project ij is undertaken and zero otherwise. If $\lambda_{ij} = 0$, the constraint set 2 limits flow over the unimproved link to its upper capacity $U_{ij}$ and constraint set 3 prohibits any flow over the improved link. If the investment project on link ij is undertaken and $\lambda_{ij} = 1$, then by constraint 2 flow over the unimproved road is prohibited and by constraint 3 flow over the improved link is limited to the upper capacity $U'_{ij}$.

Investment in network infrastructure is limited by the availability of investment funds. If $K_{ij}$ is the capital cost of the project for link ij and K is the exogenously supplied investment budget, then the constraint

$$\sum_{ij\epsilon L^*} \lambda_{ij} K_{ij} \leqslant K \tag{4}$$

limits the total capital outlay to the predetermined upper bound.

The objective of total commodity flow cost minimisation is expressed as

$$\text{MINIMISE} \sum_{g=1}^{G} \sum_{i=1}^{N} \sum_{j=1}^{N} c_{ijg} T_{ijg} + c'_{ijg} T'_{ijg} \tag{5}$$

subject to constraint sets 1, 2, 3 and 4. Alternatively, the minimisation of the investment costs necessary to sustain the given levels of supply and demand might itself be treated as an objective, perhaps subject to a constraint on the average cost of a commodity shipment. This might be a reasonable investment strategy in developing countries, characterised as they are by shortages of investment capital.

The equation and inequalities 1-5 constitute a 0-1 mixed-integer mathematical programme whose solution determines the transport cost minimising set of investment projects for a given network and the commodity flow over that enlarged network. Given the longevity of roads, network additions are synthesised for a planning horizon twenty-five to forty years in the future using point estimates of commodity supply and demand (O'Sullivan 1975). The investment projects selected for completion can then be sequenced for construction using dynamic programming and the initial capital outlays determined (see Funk and Tillman 1968, and Meyer and Straszheim 1971). Or, the model can be expressed in an intertemporal format so that the project selection and time phasing decisions are simultaneously solved (Taborga 1966, Ochoa-Rosso 1968, and Barber 1975, 1977). Using either approach this type of model is used to determine a *single* long term investment plan for the geographical evolution of the transportation network.

Consider now an alternative view of this investment problem. Suppose, more realistically, that no single point estimate of commodity supply and demand can be forecast with any degree of accuracy. Assume instead that a set of different estimates can be generated, each of which can be considered to represent a possible future state of the world. Since the forecast period normally covers the twenty-five year planning horizon and the twenty to twenty-five year lifetime of the transport investment infrastructure, this characterisation of possibility might be the best that could be hoped for realistically. When considering any investment project for implementation during the initial stages of a long-term plan it must be borne in mind that as part of the transport network it might ultimately end up operating within any one of the set of possible future operating environments. In the context of the investment model described above these futures might correspond to conditions resulting from different rates of growth of commodity supply and demand. Or, in a geographic context, the futures might correspond to different spatial distributions of commodity supply and demand, some of which might be expected if the government intervenes effectively in certain regions or sectors of the economy.

Recognising the sequential nature of the investment plan and the degree of uncertainty associated with the future operating environment of the transport network it seems most unwise to restrict the focus of the analysis to the determination of a single long-term plan. Should the future upon which the plan is based fail to materialise, it is possible that an extremely inefficient transport system may be created. Because the investment plan is carried out in stages it is theoretically possible at any point in time to revise the parts of the plan not already irrevocably committed. For example, entirely new transport links or link upgradings planned during late stages of the investment plan need not be undertaken if the future upon which their selection was based actually does not (or appears unlikely to) occur. A much shrewder strategy is to select *initial* investment projects which provide maximum flexibility in adapting to perceived changes in the evolution of the future operating environment or the social value system. Subsequent stages in the plan are then determined at later dates when the degree of uncertainty has presumably been somewhat reduced.

**Decision Strategies in the Face of Uncertainty**

Assume there are n different plausible future operating environments for a given transport network over some specific planning horizon. Denote these futures $S_1, S_2, \ldots S_n$. Now it is possible to use a conventional network investment model, such as the mathematical programme specified by equations 1-5, to determine the optimal package of investments for any single future, retaining the assumption that the preferences expressed in the composition of the optimand remain unchanged through time as does the behaviour generating its variable quantities. Obviously there is no guarantee that the optimal set of link investment projects is the same for each future and indeed there is even the possibility that no project is common to all futures. The decision criteria needed in this situation are those which can be used to select initial network investments wisely. One approach to this problem is to modify traditional uncertainty decision criteria, such as the regret or Laplace criteria, to fit the circumstances. Secondly, it is possible to use the recently developed concepts of robustness and stability in formulating efficient investment strategies.

*The Laplace Criterion*

In the absence of any *a priori* knowledge of the probability of occurrence of any future, the Laplace criterion asserts that each future can be assigned an equal probability of occurrence. The expected future is then

$$\sum_{i=1}^{n} p_i S_i$$

where $p_i$ is the probability that future i will occur,

$$\sum_{i=1}^{n} p_i = 1$$

and $p_1 = p_2, p_3 \ldots, p_n$. On the face of it, this seems sensible and a reasonable strategy would be to optimise the investment package for this expected or average future. Several objections may be raised against such a procedure. The fundamental, philosophical objection was raised in Chapter 4. The assumption of equiprobability is an unwarranted presumption of knowledge. To presume that several possible futures have an equal probability of occurrence has no basis. In this particular instance, if the futures represent different geographical distributions of transport demand then any transport investment decision based on an 'average' demand might be a poor choice in any future that could be realised. Thirdly, the costs associated with a poor investment decision need not be symmetric about this average future. Finally, in cases of great uncertainty and in situations where the decision is irreversible, a selection criterion which assumes the decision can be repeated a large number of times is irrelevant. For these reasons, the Laplace criterion appears unsuited to this problem.

*Regret Minimisation*

For each future it is possible to determine a) the optimal set of link projects and minimum level of total daily operating costs for a prescribed capital investment budget, and b) the optimal set of link projects and minimum level of capital expenditure required to meet the transport demand associated with the future. If any link investment project is selected for construction which is a component of the optimal plan for some future and that future does in fact occur, then there will be no *regret* with having initially made that decision. If, however, a project is chosen which is not a component of the optimal plan for that future then there is some regret with having made the initial decision to make that link improvement. In case a) a measure of regret is the total daily operating cost of the transport network beyond the minimum possible, and in the case of b) regret can be measured as the capital investment beyond the minimum necessary to meet transport demand. The more frequently a link appears as a component of the optimal plan in various futures, the smaller the operating cost or investment regret. Two decision rules based on regret can be formulated. First, the project with the minimum aggregate regret over all futures can be selected. Or the project with the minimum maximum regret in any single future can be chosen.

*Robustness*

Given an overall goal of selecting initial link improvements which are good investment choices in a variety of possible future operating environments for the transport system and a lack of security in the components and weight of the objective employed, it is advantageous to examine not only the package of investments which minimises this objective but also other investment plans which differ from the minimum

on the littoral, the central plains and in river valleys. Roads traverse and connect these areas. The network also reflects the market system, radiating from the ports of Makasar and Parepare to local centres which function as collectors of local produce and distributors of imported goods. The road system is in bad condition. Of the 2,188 kilometres under the jurisdiction of the provincial Public Works Department in 1970, 1,561 gravel roads, mostly in bad condition, and 685 were bituminous roads, of which 175 kilometres were of low standard. Lack of finance, the war of independence and the Darul Islam rebellion have all contributed to this situation. In the later 1960s the government, realising the relationship between a declining economy and degraded roads, began to renew most of the important links but, because of the dearth of finance and the large number of roads, little could be done. The results of a traffic count taken in 1968 are shown in Figure 8.2. The major traffic flows occur between and around Makasar and Parepare but the low volume on many links does not necessarily imply a low latent demand. Most of the roads were in bad condition at the time; indeed, 500 metres of the road from Madjene to Mamudju had been wiped out by a landslide.

Renewed interest in transport investment stems from a government wish to increase rice production in Sulawesi, one of the Indonesian provinces with a considerable potential for increased production. Commodity flow is dominated by the export flow of rice which proceeds from Wadjo, Bone and Palopo in the interior to either Makasar or Parepare. The flow of rice accounts for over ninety per cent of the region's commodity flow, which also includes maize and a set of import items including motor fuel, industrial fuel and manufactured goods.

### *Forecasts of Demand, Output and Costs*

Despite a scarcity of data for both the economy and the road system, projections of regional production and consumption as well as vehicle operating costs could be made using quite simple procedures (Ahmed et al. 1976). Because of the agrarian nature of the economy it was felt that a simple surplus-deficit model was the most realistic method of estimating freight demand. For each of the commodities transported, sub-regional consumption is forecast by relating the initial level of consumption and assumed growth rates for population and *per capita* consumption. Sub-regional production is determined by relating the initial level of production to growth rates for average yield and cultivated acreage. From both a regional as well as a national standpoint, rice is the most important crop. The major constraint to increased production is the lack of sufficient water and its adequate control. Increased output is to be achieved through the use of major irrigation schemes. In order to project the future expansion of the irrigation network, potentially irrigable land is divided into three classes. The first class consists of non-cultivated rainfed land for which new irrigation facilities are to be provided by the government. The second class of land consists of land cultivated under the traditional rainfed system and capable of transformation into a fully technical system. The final class of land consists of dry land and swampy areas with a long run potential for conversion to a technical irrigation system. The key to the forecast is an accurate estimate of the rate of conversion of these land classes. Initially it is assumed that a growth rate of above five per cent in total acreage is feasible.

Vehicle operating costs were projected using a procedure developed by the Harvard Transport Research Program and adapted to the situation of South Sulawesi (see Harral, 1970). A rather simple model is used to evaluate the vehicle performance and operating costs of different vehicle classes on the routes of specified physical characteristics. These operating cost estimates are possibly more reliable than the commodity surplus-deficit predictions if for no other reason than the fact that uncertainty in transport planning is more important on the demand as opposed to the supply side. Unpredictable increases in fuel costs, oil costs, or other operating expenses are also likely to have a region-wide impact and little, if any, geographically specific effects in South Sulawesi.

The planning period of interest commenced in 1974 coinciding with the initiation of the second Five Year Development Plan. Assuming a twenty-five year planning horizon and a twenty year lifetime beyond this horizon for the investment projects for twenty of the twenty-eight network links illustrated in Figure 8.3, forecasts are required for the year 2018. The length of the forecast period, the nature of the available data, and the simplicity of the forecasting methodology are suggestive of the high degree of uncertainty involved, particularly for the surplus-deficit estimates of freight demand.

### *Generation of Alternative Futures*

As has been demonstrated earlier, point estimates of transport demand cannot be accepted with complete certainty. For illustrative purposes, three different futures are assumed for transport demand to characterise the set of plausible futures. The first future corresponds to a realised transport demand twenty per cent less than the original projection; the

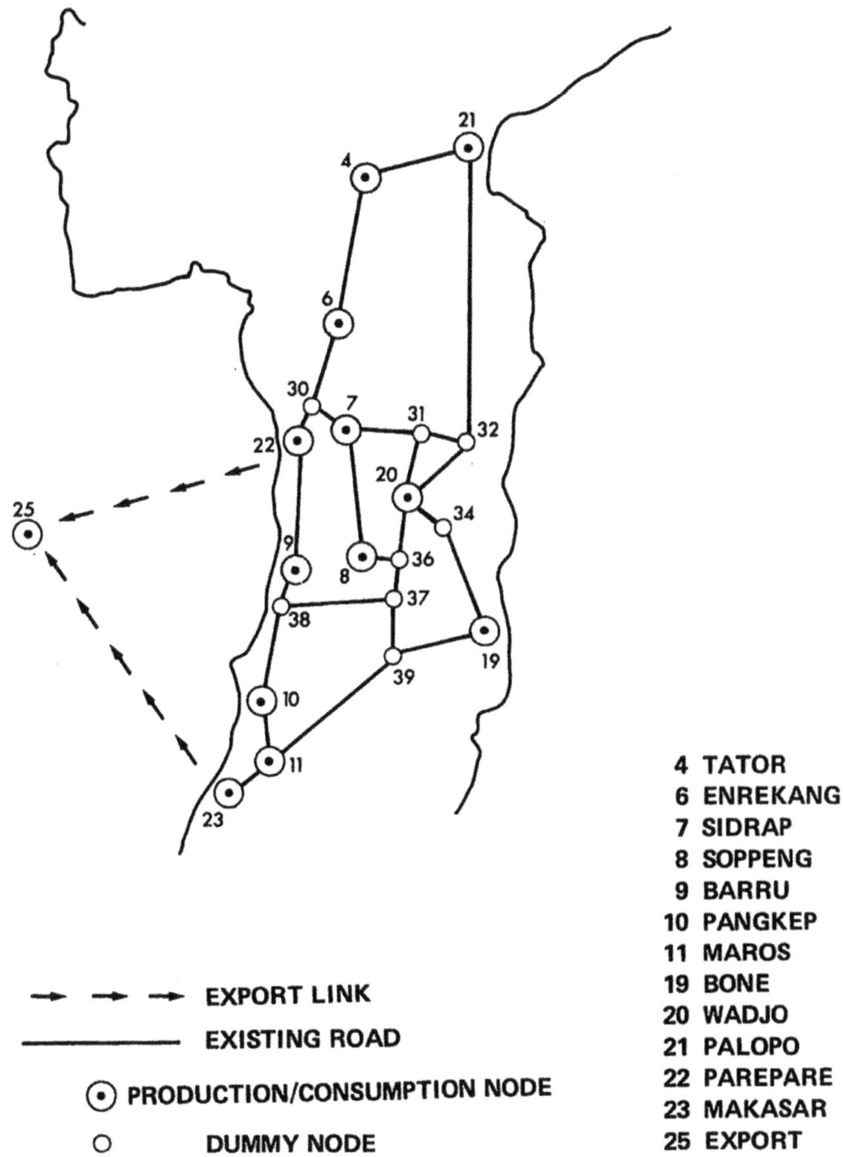


Figure 8.3: The Highway Network of South Sulawesi

second forecast corresponds to the original point estimates; and the third future corresponds to a level of transport demand twenty per cent higher than the original forecast. These three futures are denoted the low, medium, and high forecasts respectively.

Corresponding to each of these three futures is a projection of transport demand for each of four commodities: 1) rice, 2) maize, 3) import commodities, and 4) export commodities. In Figure 8.4 the projection of rice production in South Sulawesi for each of the three futures is illustrated for the period 1978-2018. Faced with the possibility that any of these three future paths of growth may actually occur, what initial link improvement project appears to be the best investment alternative?

**Figure 8.4: Alternate Projections of Rice Population in South Sulawesi 1978-2018**

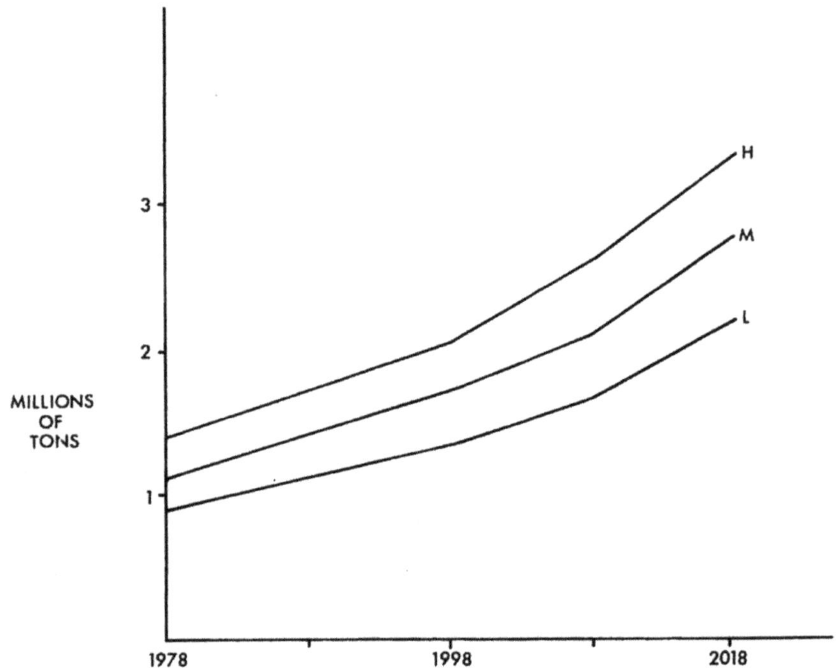

on Trunk Road Assessment, Reported personally by John Collings, February
Beiser A. 1967. *Concepts of modern physics*. New York: McGraw-Hill
Belloc H. 1923. *The road*. Manchester: Hobson
Bergendahl G. 1969. *Models for investment in a road network*. Stockholm: Bonniers
Bergendahl G. 1969. 'A combined linear and dynamic programming model for interdependent road investment planning'. *Transportation Research*, 3, 211-228
Blunden W.R. 1971. *The land use/transport system*. New York: Pergamon
Boyne G.G. 1972. 'The optimum design of fluid distribution networks with particular reference to low pressure gas distribution networks'. *International Journal for Numerical Methods in Engineering*, 5, 253-70
Brito S. 1973. 'Discussion of Dusonchet and El-Abiad'. *IEEE Transactions on Power Apparatus and Systems*, PAS-92, 1366-67
Butcher W.S., Y.Y. Haines and W.A. Hall. 1969. 'Dynamic programming for the optimal sequencing of water supply projects'. *Water Resources Research*, 5, 1196-1204
Butkovich G. and J.P. Heaney. 1975. 'Multi-objective analysis of the upper St. Johns water resource project'. Paper presented at ORSA/TIMS Joint Meeting Chicago, 30 April–2 May
Chan W.L. and G. Leininger. 1972. 'Dual characterizations of optimal systems'. *Proceedings of the 1972 IEEE Conference on Decisions and Control*. New Orleans, 466-50
Chan Y. 1972. 'Towards a more rational airline route structure: an optimization approach'. *Proceedings of the 1972 IEEE Conference on Decision and Control*. New Orleans, 506-10
Charnes, A., M. Kirby and W. Raike. 1966. 'Chance-constrained generalized networks'. *Operations Research*, 14, 1113-20
Chicago Area Transportation Study. 1956. *Forecasting economic activity: a preliminary discussion*. Chicago: CATS
Chipman J.S. and J.C. Moore. 1978. 'The scope of consumer's surplus arguments'. In A.M. Tang, F.M. Westfield and J.S. Worky (eds.) *Evolution, welfare and time in economics*. Lexington Mass: Heath
Clark C. and M. Haswell. 1964. *The economic of subsistence agriculture*. London: Macmillan
Clifford W.K. 1955. 'On the aims and instruments of scientific thought' 1872. Quoted in J.R. Newman (ed.) *The common sense of the exact sciences*. New York: Dover frontispiece

Cole L. 1965. *Planning capital investment under uncertainty with application to highway transportation systems*. PhD dissertation. Harvard University

Cole L. 1971. 'Uncertainty and facility design'. Chapter 13 of J.M. Meyer and M.R. Straszheim *Pricing and project evaluation*. Washington D.C.: Brookings

Cooley C.H. 1974. 'The theory of transportation'. *Publications of the American Economic Association*, 9, 1894. Reproduced in M.E.E. Hurst (ed.) *Transportation Geography*. New York: McGraw Hill 15-29

Cournot A.A. 1838. *Recherches sur les principes mathematiques de la theorie des richesses*. Paris: Hachette Chapter 10

Cummings R.G. 1974. *Interbasin water transfers*. Washington D.C. Resources for the Future

Dantzig G.B. 1963. *Linear programming and extensions*. Princeton: Princeton U P

David P.E. 1969. 'Transportation innovation and economic growth: Professor Fogel on and off the rails'. *Economic History Review*, Second Series 22, 3, 506-25

Davis O.A. and M.I. Kamien. 1969. 'Externalities, information and collective action'. Joint Economic Committee U.S Congress. *The analysis and evaluation of public expenditures: the P. B system Vol I* Washington D.C.: U.S. Government Printing Office

Department of the Environment. 1972a. *COBA: a method of economic appraisal of highway schemes*. London: Department of the Environment

Department of the Environment. 1972b. *Roads in England 1971*. London: HMSO

Department of the Environment. 1974. *Technical Memorandum H6/74*. London

Dodgson J.S. 1974. 'Motorway investment, industrial transport costs and sub-regional growth: a case study of the M62'. *Regional Studies*, 8, 75-91

Dommel H.W. 1975. 'Energy networks'. *Networks*, 5, 81-4

Dupuit J. 1952. 'On the measurement of the utility of public works'. *Annales des Ponts et Chaussees*. 2nd series, vol 8 (1844). Translated by R.H. Barback for *International Economic Papers*, No. 2, 83-110

Dusonchet Y.P. and A. El-Abiad. 1973. 'Transmission planning using discrete dynamic programming'. *IEEE Transactions on Power Apparatus and Systems*, PAS-92, 1358-66

Ellet Jr C. 1839. *An essay on the laws of trade*. Richmond Va: Bernard

English J.M. 1972. 'Some investment concepts in engineering systems with particular emphasis on long-range investment'. In J.M. English (ed.) *Economics of Engineering and Social Systems*. New York: Wiley 64-93

Evans, S.P. 1976. 'Derivation and analysis of some models for combining trip distribution and assignment'. *Transportation Research*, 10 37-57

Ewing D.W. 1972. 'The time dimension'. In D.W. Ewing (ed) *Long-range planning for management*. New York: Harper and Row

Faludi A. 1973. *Planning theory* London: Pergamon

Farrar D.E. 1962. *The investment decision under uncertainty*. Englewood Cliffs N J: Prentice Hall

Fenoaltea S. 1972. 'Railroads and Italian industrial growth 1861-1913'. *Explorations in Economic History*, 9, 325-351

Fishlow A. 1965. *American railroads and the transformation of the antebellum economy*. Cambridge Mass: Harvard U P

Fogel R.W. 1964. *Railroads and American economic growth*. Baltimore: Johns Hopkins U P

Forster B.A. and J. Logan. 1973. 'The welfare economics of water resource allocation over time: comment'. *Applied Economics*, 5, 61-8

Friedlaender A.F. 1965. *The Interstate Highway System: a study in public investment*. Amsterdam: North Holland

Friend J.K. and W.N. Jessop. 1969. *Local government and strategic choice*. London: Tavistock

Funk M.L. and F.A. Tillman. 1968. Optimal construction staging by dynamic programming'. *Journal of the Highway Division ASCE*, November, 255-65

Garrison W.L. and Marble D.F. 1958. 'Analysis of highway networks: a linear programming formulation'. *Proceedings of the Highway Research Board*, 38, 1-17

Garver L.L. and R.M. Sigley. 1973. 'Discussion of Dusonchet and El-Abiad'. *IEEE Transactions on Power Apparatus and Systems*, PAS-92, 1367-8

Gazis D.C. 1975. 'Transportation networks'. *Networks*, 5, 75-9

Gupta S.K. and J. Rosenhead. 1968. 'Robustness in sequential investment decisions'. *Management Science*, 15, 18-29

Harberger A.C. 1971. 'Three basic postulates for applied welfare economics: an interpretive essay'. *Journal of Economic Literature*, September, 785-97

Harral C.A. 1970. *Highway cost performance model*. IBRD Economics Department Working Paper No. 62, Washington D.C.

Harris C.C. 1974. *Regional economic effects of alternative highway systems*. Cambridge Mass: Ballinger

Harrison A.J. and S.E. Holtermann. 1973. *Economic appraisal of transport projects and urban planning objectives*. London: Department of the Environment

Harvey J.N. 'A method of network evaluation using the output of the traffic assignment process'. *Highway Research Record*, 238, 46-63

Hawke G.R. 1970. *Railways and economic growth in England and Wales 1840-1870*. Oxford: Oxford U P

Herbert J.D. and B.H. Stevens. 1960. 'A model for the distribution of residential activity in urban areas'. *Journal of Regional Science*, 2, 21-36

Heymann Jr. H. 1965. 'The objectives of transportation'. In G. Fromm (ed.) *Transport investment and economic development*. Washington D.C.: The Brookings Institution

Hicks J.R. 1956. *A revision of demand theory*. Oxford: Clarendon Press

Hill D.M., D. Brand and W.B. Hansen. 1966. 'Prototype development of statistical land-use prediction model for Greater Boston Region'. *Highway Research Record*, 114, 51-70

Hirschliefer J. 1970. *Investment, interest and capital*. Engelwood Cliffs N.J.: Prentice Hall

Hotelling H. 1938. 'The general welfare in relation to problems of taxation and of railway and utility rates'. *Econometrica*, 6, 242-69

Howe C.W. and K.W Easter. 1971. *Interbasin transfers of water*. Baltimore: Johns Hopkins U P

Hurwicz L. 1951. *Optimality criteria for decision making under ignorance*. Technical report number 370. Cowles Commission discussion paper: Statistics

Intriligator M.D. 1978. *Mathematical optimization and economic theory*. Englewood Cliffs N.J.' Prentice-Hall

Jevons W.S. 1958. *The principles of science* 1877. Reprinted New York: Dover

Kleitman D.J. 1976. 'Comments on the first two days sessions and a brief description of a gas pipeline network construction problem'. In F.S. Roberts (ed.) *Energy: mathematics and models.* Society for Industrial and Applied Mathematics

Knight F.H. 1921. *Risk, uncertainty and profit*. Boston: Houghton Mifflin

Koopmans B.O. 1940. 'The bases of probability' *Bulletin of the American Mathematical Society*. In H.E. Kyburg and H.E. Smokler (eds.) *Studies in subjective probability*. New York: Wiley 964

Kresge D.T and P.O. Roberts. 1971. *Systems analysis and simulation models*. Washington D.C.: Brookings

Kuhn H.W. and R.F. Kuenne. 1962. 'An efficient algorithm for the numerical solution of the generalized Weber problem in spatial economics'. *Journal of Regional Science*, 4, 213

Landgren G.L. and S.W. Anderson. 1973. 'Discussion of Dusonchet and El-Abiad'. *IEEE Transactions on Power Apparatus and Systems*, PAS-92, 1368

Lin S. 1975. 'Hueristic programming as an aid to network design'. *Networks*, 5, 33-43

Little J.M.D. 1950. *A critique of welfare economics*. Oxford: Clarendon Press

Loehman E. and A. Whinston. 1970. 'The welfare economics of water resource allocation over time'. *Applied Economics*, 2, 75-99

Loehman E. and A. Whinston. 1973. 'Reply to Forster and Logan'. *Applied Economics*, 5, 61-8

Lowry I. 1964. *A model of metropolis*. Santa Monica: Rand Corporation

MacKinnon J. 1978. *The ape within us*. London: Collins

MacKinnon R.D. 1968. *System flexibility with a transportation context*. PhD dissertation. Northwestern University

MacKinnon R.D. and G.M. Barber. 1976. 'Optimization models of transportation network improvement'. Department of Geography: Northwestern University, mimeo

Macrae C.D. 1969. 'A dual principle for discrete-time linear systems with economic applications'. *IEEE Transactions on Automatic Control*, AC-14, 49-52

Marglin S. 1963. *Approaches to dynamic investment planning*. Amsterdam: North Holland

Marschak J. 1968. 'Decision making economic aspects'. In D.L. Sills (ed.) *International Encyclopedia of the Social Sciences*. New York: Macmillan 42-55

Marshall A. 1920. *Principles of economics*. London: Macmillan

Martens A. and R.S Pindyck. 1972. 'An application of optimal control to investment allocation for development planning'. *Proceedings of the 1972 IEEE Conference on Decision and Control*, New Orleans, 31-6

McIntosh P.T. and D.A. Quarmby. 1970. 'Generalized costs, and the estimation of movement costs and benefits in transport planning'. *Mathematical Advisory Unit Note 179*. London: Department of the Environment

Metzer J. 1973. 'Some economic aspects of railroad development in Tsarist Russia'. *Journal of Economic History*, 33, 314-16

Midler J.L. 1970. 'Investment in network expansion under uncertainty'. *Transportation Research*, 4, 267-80

Milnor J. 1954. 'Games against nature'. In R.M. Thrall, C.H. Coombs and R.L. Davies (eds.) *Decision Processes*. New York: Wiley

Mitchell R.B. and C. Rapkin. 1954. *Urban traffic: a function of land use*. New York: Columbia U P

Morgenstern O. 1963. *On the accuracy of economic observation*. Princeton N.J.: Princeton U P

Musgrave R.A. 1959. *The theory of public finance: a study in public economy*. New York: McGraw-Hill

Neuburger H. 1971. 'User benefit in the evaluation of transport and land use plans'. *Journal of Transport Economics and Policy*, January, 52-75

Nikaido H. 1968. *Convex structures and economic theory*. New York: Academic Press

O'Brien P. 1977. *The new economic history of the railways*. London: Croom Helm

Ochoa-Rosso F. 1968. 'Optimum project addition in urban transportation networks via descriptive traffic assignment models'. *Search and Choice in Transport Systems Planning*, 5

O'Laoghaire D.T. and D.M. Himmelblau. 1974. *Optimal expansion of a water resources system*. New York: Academic Press

O'Sullivan P. 1975. 'Regional road investment planning for developing countries: an Indonesian application'. In R.F. Peel *et al. Processes in physical and human geography: Bristol essays*. London: Heinemann 339-53

Pearman A.D. 1976. 'Transport investment appraisal in the presence of uncertainty'. *Transportation Research*, 10, 331-38

Pecknold W.M. and L.A. Neumann. 1971. 'Adaptive time-staged strategies for transportation investment'. *Proceedings of the 1971 IEEE Conference on Decision and Control*, New Orleans 512-16

Pouligren L.Y. 1970. *Risk analysis in project appraisal*. Baltimore for the World Bank by Johns Hopkins U P

Promboin R.L. 1971. *Regional impact of the United States Interstate Highway System: a macroeconomic approach*. PhD dissertation. Stanford University

Putnam S.H. 1975. 'Urban land use and transportation models: a state-of-the-art summary' *Transportation Research*, 9, 186-202

Quandt R.E. 1960. 'Models of transportation and optimal network

construction'. *Journal of Regional Science*, 2, 27-45
Quandt R.E. and W.J. Baumol. 1966. 'The demand for abstract modes: theory and measurement'. *Journal of Regional Science*, 13-26
Quarmby D.A. 1967a. 'The road programme – uncertainty in investment appraisal'. *Mathematical Advisory Unit Note 25*. London: Ministry of Transport
Quarmby D.A. 1967b. 'Choice of travel mode for the journey to work'. *Journal of Transport Economics and Policy*, 1, 1-42
Reutlinger S. 1970. *Techniques for project appraisal under uncertainty*. Baltimore for the World Bank by Johns Hopkins U P
Roberts P.O. 1971. 'Selecting and staging additions to a transport network'. Chapter 15 of J.M. Meyer and M.R. Straszheim *Pricing and project evaluation*. Washington D.C.: Brookings
Rosenhead J., M. Elton and S.K. Gupta. 1972. 'Robustness and optimality as criteria for strategic decisions'. *Operational Research Quarterly*, 23, 413-31
Rostow W.W. 1960. *The stages of economic growth: a non-communist manifesto*. London: Cambridge U P
Roy B. 1971. 'Problems and methods with multiple objective functions'. *Mathematical Programming*, 1, 239-66
Sadove A.R. and G. Fromm. 1965. 'Financing transport investment'. In G. Fromm (ed.) *Transport investment and economic development*. Washington D.C.: Brookings
Samuelson P.A. 1947. *Foundations of economic analysis*. Cambridge Mass: Harvard U P 194-5
Samuelson P.A. 1952. 'Spatial price equilibrium and linear programming'. *American Economic Review*, 42, 283-303
Samuelson P.A. 1960. 'Efficient paths of capital accumulation in terms of the calculus of variation'. In K.J. Arrow, S. Karlin and P. Suppes (eds.) *Mathematical methods in the social sciences*. Stanford Calif: Stanford U P
Sauer C.D. 1941. 'Foreword to historical geography'. *Annals, Association of American Geographers*, 31, 1-24
Savage L.J. 1951. 'The theory of statistical decision'. *Journal of the American Statistical Association*, 46, 55-67
Savage L.J. 1954. *The foundations of statistics*. New York: Wiley
Schlaifer R. 1969. *Analysis of decisions under uncertainty*. New York: McGraw Hill
Schwartz J.G. 1968. *Network flow and synthesis models for transportation planning: a survey*. Department of Civil Engineering: MIT Research Report R68-49

Simon H.A. 1956. 'Rational choice and the structure of the environment'. In H.A. Simon *Models of man*. New York: Wiley

Smith N. 1966. 'Discussion of Fishburn and Murphy 1966'. In D.B. Hertz and J. Melese (eds.) *Proceedings of the Fourth International Conference on Operational Research*. London: Wiley 913

Steenbrink P.A. 1974. *Optimization of transport networks*. London: Wiley

Stowers J.R. and E.L. Kanwit. 1966. 'The use of behavioural surveys in forecasting transportation requirements'. *Highway Research Record*, 1106, 44-51

Strotz R.H. 1965. 'Urban transportation parables'. In J. Margolis (ed) *The public economy of urban communities*. Washington D.C.: Resources for the Future

Suppes P. and J.L. Zinnes. 1963. 'Basic measurement theory'. From R.D. Luce, R.R. Bush and E. Galanter (eds.) *Handbook of Mathematical Psychology*. New York: Wiley

Taborga P. 1966. *A model to study an optimal transportation policy in Chile*. Research report R66-8, MIT Department of Civil Engineering

Takayama T. and G.G. Judge. 1971. *Spatial and temporal price and allocation models*. Amsterdam: North Holland

Theil H. 1961. *Economic forecasts and policy*. Amsterdam: North Holland

Thomas E.N. and J.L. Schofer. 1970. *Strategies for the evaluation of alternative transportation plans*. National Cooperative Highway Research Program Report 96, Washington D.C.: Highway Research Board

Tindall J. 1974. *RUCL – a road user cost program for London*. Greater London Council Department of Planning and Transportation, Research Memorandum

Townsend, C.H., H.S. Jones and A. Cleveland. 1973. *The overall design of the natural gas system of the United Kingdom*. London: International Gas Union

U.K. Parliamentary Papers 1837-1838. *Two reports of Commissioners appointed to consider and recommend a General System of Railways for Ireland*. London: HMSO, two parts, 6333

Van der Tak H.G. and A. Ray. 1971. *The economic benefits of road transport projects*. Baltimore for the International Bank for Reconstruction and Development by Johns Hopkins U P

von Neumann J. and O. Morgenstern. 1944. *Theory of games and economic behaviour*. Princeton N.J.: Wiley

Voorhees A. 1955. 'A general theory of traffic movement'. *Proceedings*

*of the Institute of Traffic Engineers*. 46-56
Vuchic V.R. 1971. 'Concepts of flexibility in transportation systems analysis'. *High Speed Ground Transportation Journal*, 5, 53-62
Walters A.A. 1968. *The economics of road user changes*. Baltimore for the International Bank for Reconstruction and Development by Johns Hopkins U P
Wang J., R.R. Snell and M.L. Funk. 1968. 'Towards a solution for the optimal allocation of investment in urban transportation networks'. *Highway Research Record*, 238, 23-45
Warner S.L. 1962. *Stochastic choice of mode in urban travel: a study in binary choice*. Evanston Illinois Transportation Center: Northwestern U P
Weber A. 1909. *Uber den Standort den Industrien*. Tubingen: Mohr
Weisskopf V.F. 1972. 'Quality and quantity in quantum physics'. In V.F. Weisskopf *Physics in the twentieth century: selected essays*. Cambridge Mass: MIT
Wellington A.M. 1887. *The economic theory of the location of railways*. New York: Wiley
Willig R. 1976. 'Consumer's surplus without apology'. *The American Economic Review*, 66, 589-97
Wilson A.G. 1967. 'A statistical theory of spatial distribution models'. *Transportation Research*, 1, 253-69
Wilson A.G. 1970. *Entropy in urban and regional modelling*. London: Pion
Winsten C.B. 1968. 'Regression analysis versus category analysis'. *Proceedings of the PTRC Seminar on Trip End Estimation*. London: PTCR 17-19
Zadeh L. 1965. 'Fuzzy sets'. *Information and Control*, 8, 338-53

# INDEX